**Electronics texts for engineers and scientists**

Editors: H. Ahmed and P. J. Spreadbury
*Lecturers in Engineering, University of Cambridge*

# Electric and magnetic fields

## An introduction

# Electric and magnetic fields

## An introduction

SIR CHARLES OATLEY, OBE, FRS
*Emeritus Professor of Electrical Engineering*
*University of Cambridge*

CAMBRIDGE UNIVERSITY PRESS
CAMBRIDGE
LONDON NEW YORK MELBOURNE

Published by the Syndics of the Cambridge University Press
The Pitt Building, Trumpington Street, Cambridge CB2 1RP
Bentley House, 200 Euston Road, London NW1 2DB
32 East 57th Street, New York, NY 10022, USA
296 Beaconsfield Parade, Middle Park, Melbourne 3206, Australia



First published 1976

Printed in Great Britain at the
University Printing House, Cambridge
(Harry Myers, University Printer)

**Library of Congress Cataloguing in Publication Data**

Oatley, Sir Charles William

Electric and magnetic fields

(Electronics texts for engineers and scientists)

Includes index

1. Electromagnetic fields  2. Electromagnetic theory  I. Title

QC665.E402  530.1′41  76–7137

ISBN 0 521 21228 6 hard covers
ISBN 0 521 29076 7 paperback

# Contents

## 4 The magnetic field in free space



## 5 Electric and magnetic fields in material media



## 6 Methods of solution when $\sigma$, $\epsilon$ and $\mu$ are constant



## 7 Non-linear materials

# Preface

In many universities, undergraduates who are reading engineering attend, during their first or second year, an introductory course in electromagnetic theory, though many of them will later be specializing in a non-electrical branch of engineering. A similar situation arises with undergraduates reading natural sciences, who include physics at the outset, but will later specialize in some other science. The present book is based on lectures which I gave for many years, for a course of this kind, in the Cambridge University Engineering Department.

Almost everyone attending a university or polytechnic degree course will already have some knowledge of electromagnetics, but the instruction previously received is unlikely to have been very systematic and the undergraduate may well be uncertain about the meanings of quite elementary concepts such as flux vectors, line integrals, or even solid angles, On the other hand he will be quick to detect any illogicality of presentation and will resent the introduction of statements of fact which have received no proper justification. He may, or may not, be familiar with vector algebra, but is unlikely to have encountered vector calculus. The course thus brings him face to face with a considerable number of new ideas and his difficulties are more likely to be conceptual than mathematical. Unless he is led forward by easy stages, he may well concentrate on a few rules that will enable him to answer the limited range of questions that an examiner can set, without any real understanding of the subject.

Bearing these facts in mind, I have tried to write a book in which the emphasis is on physical principles and the mathematics has been kept to a minimum. At the same time, I hope that the reader will gain an overall picture of the more advanced mathematical methods that are available, should he need them. More space than is usual has been devoted to a discussion of the meaning of electric and magnetic fields inside a material medium and the treatment is designed to apply to non-isotropic, non-linear media, as well as to those for which permittivity and permeability are constants. The problems at the ends of the chapters are mostly straightforward and are chosen to test knowledge of principles rather than mathematical dexterity.

Since this book is intended to cover the content of a non-specialist introductory course, it contains no treatment of aerials, waveguides and related topics. Undergraduates who intend to specialize in electrical engineering or physics require a much more detailed discussion of these subjects than could properly be included in a first course, and it is intended to cater for their needs in another book in this series.

The inclusion of a brief account of the motion of charged particles in electric and magnetic fields can, I believe, be justified. Engineers and scientists of all kinds may later need to use oscilloscopes, electron microscopes, mass spectrometers and so forth, and should have sufficient knowledge of fundamentals to enable them to understand the principles underlying the operation of such instruments. Time is unlikely to be found for a separate course on electron optics and the short treatment that I have given seems to fit in better with electromagnetic theory than with any other subject.

It goes without saying that I have derived great benefit from reading other books on electromagnetic theory and I hereby acknowledge my debt to authors too numerous to name.

C. W. Oatley

Cambridge, 1975

# 1

# Introduction

## 1.1 Field problems

In the elementary study of electrical science it is usual to consider the flow of current in simple circuits consisting of resistance coils, meters, bridge wires and so forth. The conductors in such circuits generally have one property in common: the length of each is large in comparison with the dimensions of its cross-section. It is then a reasonable assumption, at low frequencies, that the current is uniformly distributed over the cross-section of each conductor, and concepts such as resistivity present no difficulty.

A different type of problem is illustrated in fig. 1.1. Here we have a large rectangular block of conducting material and current $I$ is led into one face and out of the opposite face by wires whose diameters are small in comparison with the length of an edge of the cube. Within the cube, the current will clearly spread out, but it is by no means easy to see exactly what form the spreading-out will take. For a complete solution of the problem we should wish to know what fraction of the current $I$ flows through each element of the cube and in what direction the flow takes place. Given the resistivity of the material, we might also wish to know the effective resistance between the points at which the current enters and leaves the cube.

This is an example of a *field problem* and its solution requires techniques which are quite different from those used in circuit analysis. We shall find that these techniques are very similar to those needed for the solution of problems relating to electrostatic fields or magnetic fields, so it is convenient to deal with the three types of field within a single book. Incidentally, similar techniques are needed in other branches of engineering to solve problems relating to the flow of heat or of fluids, the stress in structural members or foundations and so forth.

We shall use the word *field* in a general sense to indicate the region of space throughout which the effect that we are studying is appreciable. This region may be bounded, as in the problem of fig. 1.1, where current flows only in that region of space occupied by conductors. On the other hand, the electrostatic field caused by an electric charge may extend to infinity in all directions. In this case, however, we find that any effect resulting from the

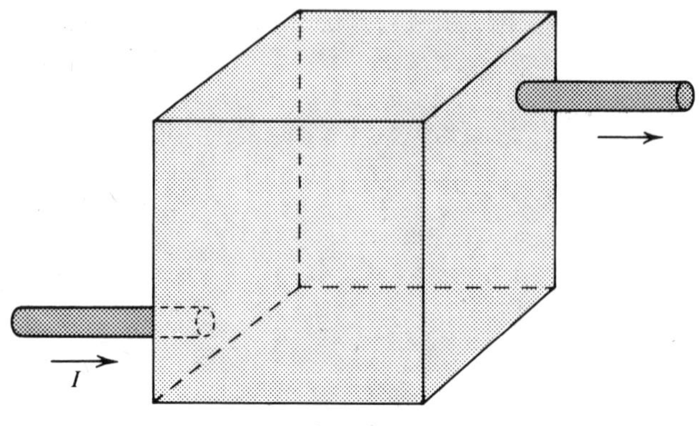


Fig. 1.1

charge becomes less apparent as we move further from the charge. We may express this fact by giving a quantitative significance to the word field, and by saying that the strength of the field decreases with distance from the charge. Clearly, for each type of field, we shall need to define some quantity whose magnitude at any point can be taken as a measure of the field strength at that point.

## 1.2 The order in which subjects will be discussed

In the next chapter we shall deal with the general problem of the flow of current in a single homogeneous isotropic medium and shall derive the basic equations governing this flow. Unless the system that we are considering possesses planar, cylindrical or spherical symmetry, the solution of these equations is likely to be quite complicated but, in the problems at the end of the chapter, there are examples where the reader can readily obtain the required answers by the use of simple mathematics.

The same plan will be followed in chapters 3 and 4, which will deal respectively with the electrostatic field in free space and with the magnetic field in free space. The latter will also be concerned with changing magnetic fields and with electromagnetic induction. Once again the primary aim will be to establish the basic equations and to illustrate them by using them to solve simple problems.

In each of the above chapters we have been dealing with only one medium: with the flow of current in a single medium, or with an electrostatic field or a magnetic field in free space. In chapter 5 we consider the complications which arise when more than one medium is present in the field. For all three types of field it will clearly be necessary to set up equations governing the conditions which obtain at the surface of separation between two media. In addition, in the cases of electrostatic and magnetic fields, we shall have to enquire whether the equations established for free space are still valid when material media are present. This question, in turn, will lead us to consider precisely what we mean by the electrostatic or magnetic field inside a material medium.

We have already said that, in the early chapters, only the simplest problems will be dealt with, so that mathematical complexity can be avoided. In chapter 6 we give a brief account of the general methods available for the solution of more difficult problems and we use these methods to obtain results of practical importance. It is convenient to consider all three types of field in the same chapter since, to a great extent, the same mathematical techniques are needed for all of them.

The theory presented in these first six chapters can be applied only to materials of a restricted class. They must be homogeneous and isotropic and, in addition, they must be *linear*. In the case of a medium through which current is flowing, this means that the material must obey Ohm's law, so that the resistivity is a constant. For electrostatic and magnetic fields we shall find that there are corresponding quantities, known as the *permittivity* and the *permeability* of the medium, and linearity requires that these also shall be constants. However, many materials of great technological importance do not conform to these restrictions and chapter 7 is devoted to a discussion of the methods that must be employed when dealing with these substances.

In chapter 8 we describe the storage of energy in electrostatic and magnetic fields. This leads to methods whereby the forces on bodies situated in electrostatic or magnetic fields can be calculated. It also enables us to prove certain general theorems of wide application.

It was pointed out by Maxwell that the theory developed in the above chapters leads to an anomaly when applied to an electric field which is changing with time. In chapter 9 we discuss this matter, and give an account of the hypothesis which Maxwell introduced to remove the anomaly and which led him to predict the existence of electromagnetic waves.

Finally, in chapter 10, we deal with four topics of importance, which do not fit readily into the earlier chapters.

## 1.3 Vector notation

From what has been said, it will be clear that field problems are usually three-dimensional problems and, for this reason, it is convenient to use vector notation when dealing with them. It must, however, be emphasized that this is a convenience rather than a necessity, since almost the whole of the theory with which we shall be concerned was developed before the use of vector notation became common. Vector algebra and vector calculus do not enable us to derive any results that could not have been obtained without their aid, but they do make it possible to carry out the mathematical manipulations much more concisely and to express the results in particularly simple forms.

Since the reader may not have had much previous experience of the use of vector methods the various vector operations will be introduced gradually, as and when they are needed, and the meaning of each will be fully explained.

## 1.4 The establishment of a physical theory

A physical theory must be based on a number of postulates, which are assumed to be true. One of the difficulties with electromagnetic theory is that different writers, at different times, have adopted quite different sets of basic postulates, so that one author is at pains to prove something which another accepts as an axiom. About the final theory there is no disagreement, so it is only necessary for us to state the procedure to be followed in this book.

The truth of the basic postulates must rest ultimately on experimental evidence and, following the historical development of the subject, we shall normally choose those postulates that can be most directly related to the simple experiments carried out by the early pioneers who laid the foundations of the science of electromagnetism. However, to take account of more recent work, it will sometimes be appropriate to state the postulates in forms differing somewhat from their original formulations.

The pioneers had necessarily to work with crude apparatus and their experiments could not have been performed with any great accuracy. At the outset, therefore, our acceptance of the truth of the laws which they enunciated will be an act of faith. At a later stage we shall find indirect consequences of the laws which can be verified with the highest accuracy (§10.1).

## 1.5 Macroscopic and microscopic theories

It has already been stated that we shall have to consider how the presence of material substances can affect electrostatic and magnetic fields. Unless the contrary is specifically stated we shall assume these materials to be both homogeneous and isotropic, though neither of these assumptions is really true.

Table 1.1

| Quantity | Unit | Symbol | Section reference |
|---|---|---|---|
| Electric capacitance | farad | F | 3.7.1 |
| Electric charge | coulomb | C | 2.1.1 |
| Electric conductance | siemens | S | 2.1.2 |
| Electric conductivity | siemens metre$^{-1}$ | S m$^{-1}$ | 2.5.1 |
| Electric current | ampere | A | 2.1.1 |
| Electric current density | ampere metre$^{-2}$ | A m$^{-2}$ | 2.1.3 |
| Electric displacement | coulomb metre$^{-2}$ | C m$^{-2}$ | 3.4 |
| Electric field strength | volt metre$^{-1}$ | V m$^{-1}$ | 2.5.1 and 3.2.1 |
| Electric potential difference | volt | V | 2.1.1 |
| Electric resistance | ohm | Ω | 2.1.2 |
| Electric resistivity | ohm metre | Ω m | 2.1.2 |
| Electromotive force | volt | V | 4.4.2 |
| Energy | joule | J | 2.1.1 |
| Magnetic field strength | ampere metre$^{-1}$ | A m$^{-1}$ | 4.1.5 |
| Magnetic flux | weber | Wb | 4.1.4 |
| Magnetic flux density | tesla | T | 4.1.2 |
| Magnetic scalar potential | ampere | A | 4.1.6 |
| Magnetomotive force | ampere | A | 7.2.7 |
| Mutual inductance | henry | H | 4.5.1 |
| Permeability of free space | henry metre$^{-1}$ | H m$^{-1}$ | 4.1.2 |
| Permittivity of free space | farad metre$^{-1}$ | F m$^{-1}$ | 3.1.2 |
| Power | watt | W | 2.1.1 |
| Self-inductance | henry | H | 4.5.3 |

Table 1.2

| Multiple | Prefix | Symbol | Multiple | Prefix | Symbol |
|---|---|---|---|---|---|
| $10^{-1}$ | deci | d | 10 | deca | da |
| $10^{-2}$ | centi | c | $10^{2}$ | hecto | h |
| $10^{-3}$ | milli | m | $10^{3}$ | kilo | k |
| $10^{-6}$ | micro | $\mu$ | $10^{6}$ | mega | M |
| $10^{-9}$ | nano | n | $10^{9}$ | giga | G |
| $10^{-12}$ | pico | p | $10^{12}$ | tera | T |
| $10^{-15}$ | femto | f | | | |
| $10^{-18}$ | atto | a | | | |

A homogeneous material is one in which every particle, no matter how small, is exactly the same as every other particle and this can never be true on a sub-atomic scale. Implicitly, therefore, we are limiting the application of our theory to pieces of material that are large in comparison with the dimensions of an atom.

An isotropic material is one which has exactly the same properties in all directions. Even for a perfect single crystal this would not be the case unless the material crystallized in the cubic system, and the materials to which our theory must apply will not normally be single crystals. They will, however, very commonly be composed of minute crystallites whose axes are orientated at random so that, for a piece of the material which is large in comparison with a crystallite, the average properties will be the same in all directions. Once again, we are saying that our theory will be valid only if the pieces of material are not too small.

Summing up, the theory that we propose to develop is essentially a macroscopic theory and the properties of materials that we shall use are those obtained by measurements on specimens that are large compared with the dimensions of atoms or crystallites. It will be inherently impossible for our theory to provide any information about phenomena on an atomic scale. This is in line with the fact that classical electromagnetic theory was developed long before the advent of quantum theory.

## 1.6 Units and symbols

Our theory will require the introduction of a number of new units and other quantities. For easy reference, a list of these is given in table 1.1, together with the appropriate symbol for each quantity and a reference to the section in which it is first introduced. Similarly, table 1.2 lists the prefixes that are used to construct multiplies and sub-multiples of the units.

# 2

# The flow of current in a homogenous isotropic medium

## 2.1 Statement of the problem

### 2.1.1 Units

We assume, as an experimental fact, that the passage of current through a medium results from the flow of charge under the action of an applied potential difference. Thus, at the outset, we need definitions of the units of these three electrical quantities.

Throughout this book we shall use the International System of Units (SI) and the reader is assumed to be familiar with the mechanical units in this system. The basic electrical unit is the *ampere* (symbol A) and is defined as follows: *the ampere is that constant current which, if maintained in two straight parallel conductors of infinite length, of negligible circular cross-section, and placed one metre apart in vacuum, would produce between these conductors a force equal to* $2 \times 10^{-7}$ *newton per metre of length.*

The significance of this definition and the means whereby it leads to a practical realization of the unit of current will be explained at a later stage (§4.3.2, §10.2.2).

The unit of electric charge is the *coulomb* (symbol C). It is defined to be the quantity of electricity carried, in one second, past any cross-section of a circuit in which an unvarying current of one ampere is flowing.

We know from experiment that a continuous current will not flow through a conductor unless a difference of potential is maintained between the ends of the conductors. This may be done by connecting the conductor to a battery, a generator, or some similar device. The unit of potential difference is the *volt* (symbol V). It is defined by the statement that energy of 1 *joule* (symbol J) is needed to convey one coulomb of positive electric charge from one point to another, when the potential of the second point is one volt higher than that of the first. (N.B. we shall consider later (§4.4.2) the rather different situation that arises when current in a closed circuit is caused by electromagnetic induction.)

It follows from the above definitions that when current $I$ amperes flows from a given point to a second point whose potential is $V$ volts lower than that of the first, energy is dissipated at the rate of $VI$ joules second$^{-1}$, or $VI$ *watts* (symbol W). This energy usually appears as heat.

Other units will be introduced and defined when the need arises and, for reference, a complete list is given in table 1.1 (p. 5).

### 2.1.2 Assumptions

Throughout this chapter we shall assume that the medium through which current is flowing is both homogeneous and isotropic, on a macroscopic scale (§1.5). We shall also assume that any conductor made of the material obeys Ohm's law. This means that if a potential difference $V$ is maintained between two areas of the conductor (each area having the same potential throughout) the resulting current $I$ flowing between these areas is proportional to $V$. Thus, we may write

$$V/I = \text{constant} = R \tag{2.1}$$

The constant $R$ is known as the *resistance* of the conductor between the specified areas and, when $V$ is measured in volts and $I$ in amperes, $R$ is expressed in *ohms* (symbol $\Omega$). The reciprocal of the resistance is the *conductance* expressed in *siemens* (symbol S).

For a homogeneous isotropic material (2.1) implies that the resistance of a rod of constant cross-section, in which current enters uniformly over one end and leaves uniformly over the other, can be written in the form

$$\text{resistance} = \frac{\rho \times \text{length}}{\text{area of cross-section}} \tag{2.2}$$

where $\rho$ is a constant which is known as the *resistivity* of the material. The SI unit for resistivity is the *ohm metre* ($\Omega$ m).

Let us consider what experimental work would be needed to determine whether the above assumptions were valid for some new material, of which we had been given a lump of reasonable size. From the lump we might cut a rod of uniform cross-section, whose length was large compared with its lateral dimensions. If current from a battery enters at one end and leaves at the other end of the rod, and if the material is homogeneous, we may expect the current to be distributed uniformly over the cross-section, except at points near the ends. The distribution of potential at points remote from the ends can be investigated, with an instrument such as a potentiometer, which absorbs no current. From a number of such measurements, with different measured values of current, we can ascertain whether Ohm's law is obeyed and, if so, a value for the resistivity can be determined. If a number of such measurements, carried out on rods of differing lengths and cross-sections and cut from the original lump of material with differing orientations, all give the same value for the resistivity, we may have faith that the material is both homogeneous and isotropic. The point to be made is that there is no unique experiment which will tell us whether the assumptions that we are making are valid for a particular material; the properties of new materials emerge gradually, as the result of a considerable number of measurements.

A great many materials do not conform to the assumptions that we have made. Ohm's law may not be obeyed; the material may not be isotropic, particularly if it has been formed by rolling or extrusion; if the resistivity is very large, surface conductivity may be comparable with, or greater than, conductivity through the bulk of the material. The theory which we shall develop in this chapter cannot be applied to such materials without further consideration.

There is one further limitation that we shall place on our theory. We shall assume that current is carried by the motion of charged particles of a single sign. In many conductors charges of opposite sign are effectively present simultaneously; holes and electrons in a semiconductor or positive and negative particles in a gaseous discharge. It is not implied that electromagnetic theory is useless in such cases; merely that the complications resulting from the recombination or spontaneous generation of charges are outside the scope of this book.

### 2.1.3 The nature of the required solution

Let us suppose that current enters a conducting medium through a small area, which we shall term the *source*, and leaves through another small area, which will be termed the *sink*. To simplify matters, we suppose the potential over the whole of the source to be constant and similarly for the sink. (If this were not true, it would be necessary to suppose the medium to be supplied with current from a finite number or from a continuous distribution of sources and sinks, of differing potentials, but the theory which we shall develop will include such cases.)

At the outset, therefore, we shall be given the constant potential difference between source and sink and the solution of our problem must include information about the way in which the potential varies from point to point of the intervening medium. Since the potential at a point is represented by a scalar number, a convenient method of representing the distribution of potential is to construct a series of surfaces joining points which are at the same potential. Such surfaces are known as *equipotential surfaces* and it is usually convenient to construct them for equal steps of potential.

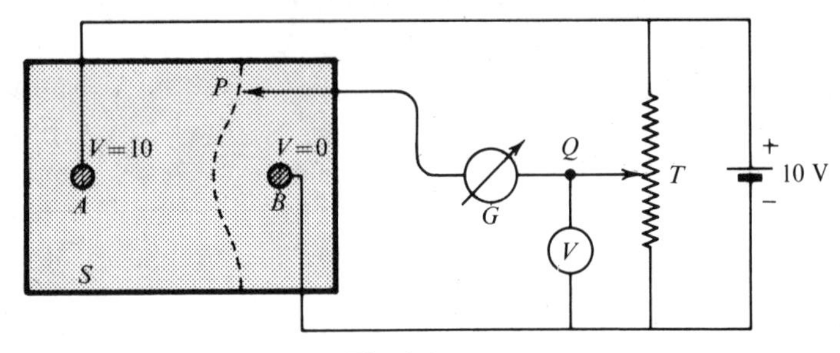


Fig. 2.1

Although we shall ultimately be dealing with a three-dimensional medium, it is instructive to consider first the two-dimensional case where the medium is a conducting sheet of uniform small thickness. The equipotential surfaces then become equipotential lines and their form can readily be determined by experiment and plotted on a plane diagram. In fig. 2.1, $S$ is a rectangular conducting sheet of uniform thickness, to which thicker studs $A$ and $B$, of highly conducting material have been soldered. $A$ and $B$ are connected to the terminals of a 10 V battery and they form the source and sink respectively. Since we shall be concerned only with differences of potential, it is immaterial what point we take as our zero and, following a convention that is often adopted, we choose the negative terminal of the battery. Then $B$ is at zero potential, while $A$ is at +10 V. Suppose we now wish to determine the 1 V equipotential. We adjust the potential divider $T$ until the voltmeter $V$ reads 1 V, which is then the potential of point $Q$. The probe $P$ is then placed in contact with the conducting sheet and a position found such that the galvanometer $G$ shows no deflection, indicating that this point lies on the 1 V equipotential. By trial, a succession of such points is found and the course of the equipotential is plotted. The voltmeter is then set to a different value and the process repeated. In this way we can plot as many equipotentials as we wish.

In the more general three-dimensional case, we cannot determine the equipotential surfaces by experiment unless the conducting medium is a liquid: even then the process is an extremely tedious one. Moreover, we cannot represent a three-dimensional potential variation on a plane diagram. In the general case there is no simple solution to this difficulty but, if the medium has one or more planes of symmetry, it may be sufficient to plot the lines in which the equipotential surfaces cut these planes.

Because of the obvious limitations of the experimental method of determining equipotentials, it must clearly be one aim of our theory to derive expressions from which the forms of the equipotentials may be calculated, either by direct solution of the equations, or by numerical computation.

Turning now to the distribution of current throughout the medium, we note that current, unlike potential, has direction as well as magnitude; it is a vector quantity. In any context where both magnitude and direction are relevant we shall represent current by the bold-faced symbol $\boldsymbol{I}$. The symbol $I$ will, however, be used when we wish to refer only to the magnitude of the current and its direction is either unimportant or has already been specified in some other way. Thus, we might write that a current $I$ is flowing in a circuit, it being understood that the direction of the current at each point is the direction of the circuit. Again, we might write that the rectangular components of the current $\boldsymbol{I}$ flowing through a particular area of a medium are $I_x$, $I_y$ and $I_z$, the directions of these cqmponents being specified by the axes. We shall use this general procedure for many other vector quantities that we shall encounter.

As soon as we begin to consider the way in which the distribution of current in a medium might be represented, it becomes clear that current is, in fact, an unsatisfactory quantity with which to work. The current at any point is necessarily zero, since finite current can only flow through an area of finite size. We therefore turn our attentinn to the *current density*, which is defined as follows. At any point $P$ take a small area $\delta S$ lying in a plane which is normal to the direction of flow of current. Let $\delta I$ be the magnitude of the current through $\delta S$. Then the magnitude of the current density $J$ at $P$ is given by

$$J = \mathrm{Lt}_{\delta S \to 0} \frac{\delta I}{\delta S} \tag{2.3}$$

The current density is clearly a vector quantity having, at every point, the same direction as $\boldsymbol{I}$. We shall represent it by $\boldsymbol{J}$. It has a definite value at every point of the medium and, since it is a vector, we need some means of representing both its magnitude and direction.

The unit of $\boldsymbol{J}$ is the *ampere metre*$^{-2}$ (A m$^{-2}$).

## 2.2 The graphical representation of $\boldsymbol{J}$

### 2.2.1 The direction of $\boldsymbol{J}$

We return to the system depicted in fig. 2.1, in which current from a source $A$ flows through a uniform conducting sheet to a sink $B$. At every point of the sheet there is a single unique direction in which current is flowing so that, starting from a point on $A$, we may trace a continuous curve such that, at every point of its length, a tangent to the curve is always in the direction of flow. Since all current from $A$ eventually reaches $B$, our curve will end at some point on $B$. We shall term such a curve a *flow line*. There is no limit to the number of flow lines that can be drawn and a few of them are shown

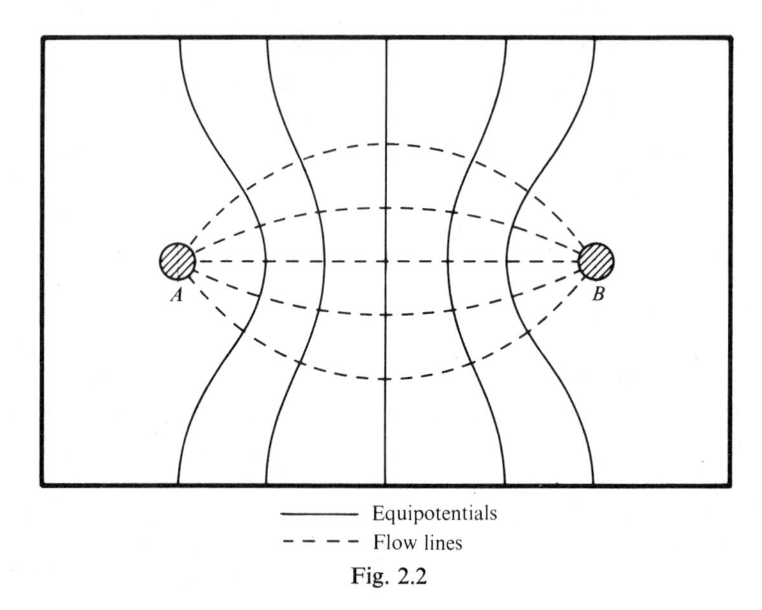


Fig. 2.2

dashed in fig. 2.2. In the same figure, a selection of equipotentials is indicated by full lines.

Since the direction of flow at any point is unique, flow lines clearly cannot cross one another. There is a second, more important property of the lines which we can deduce by considering the intersection of a flow line with an equipotential. Unless the flow line cuts the equipotential at right angles, there will be a component of current flowing along the equipotential. The existence of such a component would imply, by Ohm's law, a variation of potential along the equipotential and this, by definition, does not occur. We thus conclude that flow lines and equipotentials always cut each other at right angles.

For simplicity the above argument has been presented in terms of a two-dimensional flow of current in a uniform sheet. Clearly, however, it is equally applicable to the more general case of three-dimensional flow, where we should conclude that flow lines always intersect equipotential surfaces at right angles.

### 2.2.2 The magnitude of $J$

Consideration of fig. 2.2 shows that the flow lines are crowded together in the vicinity of $A$ and $B$, where we should expect the current density to be great, and are more widely spaced in regions where the current density

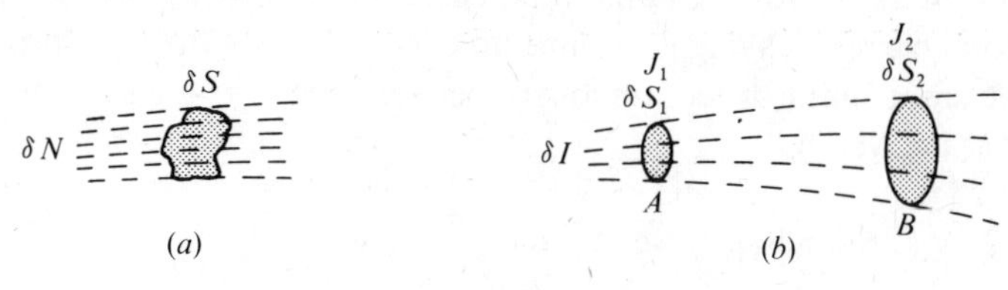


Fig. 2.3

seems likely to be small. This suggests that it might be possible to represent the magnitude of $\boldsymbol{J}$ by the density of flow lines.

We can certainly do this in any one small region of the medium, because the number of flow lines that we draw is entirely at our disposal. Thus, in fig. 2.3(*a*), let $\delta S$ be a small element of area at right angles to the direction of current flow and let the magnitude of the current density over this area be $J$. Then we can choose the number of lines $\delta N$ which pass through $\delta S$ so that

$$\mathrm{Lt}_{\delta S \to 0} \frac{\delta N}{\delta S} = J \tag{2.4}$$

in some convenient units. For example, we might arbitrarily decide to let a current density of 1 A $\mathrm{m}^{-2}$ be represented by 1000 lines per square metre. As $\delta S$ gets smaller, $\delta N$ will not usually be an integer, but this need cause no conceptual difficulty, since we may suppose the space between unit flow lines to be sub-divided by 'decilines', 'centilines' and so on. Thus we have a satisfactory method of representing the magnitude of the current density at $\delta S$ by the density of the flow lines.

However, having fixed $\delta N$ in accordance with (2.4), we have settled the density of this particular group of lines throughout the whole of their course from source to sink. It is not immediately obvious that, at all other points, they will represent by their density (i.e. by their number per unit area normal to the direction of flow) the current densities at these points. We now prove that, in fact, they will do so.

In fig. 2.3(*b*), let a particular group of flow lines represent the current density $J_1$ at a small area $\delta S_1$ at $A$, and let the same lines subsequently fill an area $\delta S_2$ at $B$, both $\delta S_1$ and $\delta S_2$ being normal to the directions of flow at $A$ and $B$ respectively. Let us arbitrarily take $N$ lines per unit area to represent unit current density. For this representation the number of lines $\delta N_1$ that must pass through $\delta S_1$ at $A$ is

$$\delta N_1 = J_1 N \delta S_1 \tag{2.5}$$

while the number passing through $\delta S_2$ at $B$ is

$$\delta N_2 = J_2 N \delta S_2 \tag{2.6}$$

However, in the steady state, the total current entering $\delta S_1$ must be equal to the total current leaving $\delta S_2$, since no current crosses the flow lines and charge cannot build up continuously between the two areas. Thus, the current $\delta I$ is given by

$$\delta I = J_1 \delta S_1 = J_2 \delta S_2 \tag{2.7}$$

From (2.5), (2.6) and (2.7) we see that

$$\delta N_1 = \delta N_2 \tag{2.8}$$

Thus the number of flow lines which, by our convention, is necessary to represent the current density at $A$ proves also to be exactly the right number to represent the current density at $B$, or at any other cross-section of the group of flow lines.

We have now shown that the current density $\boldsymbol{J}$ can be represented in both direction and magnitude by a set of flow lines. A vector such as $\boldsymbol{J}$ for which this is possible has special properties and is known as a *flux vector*. It is given this name because its properties arise from its association with the *flow* of some indestructible quantity which, in the case of $\boldsymbol{J}$, is electric charge. By no means all of the vectors that we shall encounter in field theory are flux vectors.

In practice, the representation of $\boldsymbol{J}$ in magnitude as well as direction by flow lines (or, as we shall often term them, *flux lines*), is qualitatively useful in giving us a picture of the flow of current in a conductor. Quantitatively it is strictly limited by the impossibility of representing a three-dimensional situation on a plane diagram. We have dealt with the subject at some length because the discussion may perhaps assist the reader to grasp the significance of a flux vector.

We shall shortly wish to express some of the foregoing ideas in mathematical languages but, before doing this, it is convenient to introduce some of the notation and results of vector algebra.

## 2.3 Vector algebra

### 2.3.1 The representation of vector quantities

Vector quantities, such as force, velocity or current density, which have direction as well as magnitude, are represented geometrically by straight lines. The length of the line indicates the magnitude of the quantity on some arbitrarily chosen scale, while the direction of the line, with an arrow added, tells us the direction in which the quantity is acting. In fig. 2.4(*a*) the vectors $\boldsymbol{A}$ and $-\boldsymbol{A}$ are represented.

The sum of two vectors $\boldsymbol{A}$ and $\boldsymbol{B}$ is obtained by the usual parallelogram process and the resultant is a vector $\boldsymbol{C}$, which also has direction as well as

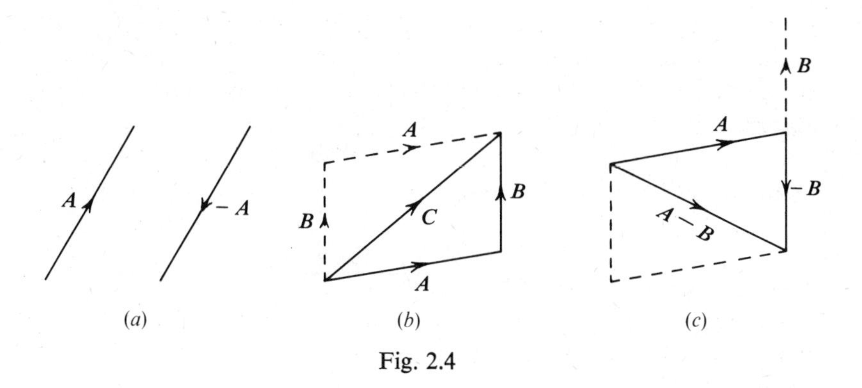


Fig. 2.4

magnitude. The process is expressed algebraically by

$$\boldsymbol{A}+\boldsymbol{B} = \boldsymbol{C} \tag{2.9}$$

and is illustrated in fig. 2.4(*b*).

To subtract $\boldsymbol{B}$ from $\boldsymbol{A}$, the direction of $\boldsymbol{B}$ is reversed, to give $-\boldsymbol{B}$, which is then added to $\boldsymbol{A}$, as shown in fig. 2.4(*c*).

It is clear from the figure that the order in which the vectors are drawn is immaterial, so that

$$\boldsymbol{A}+\boldsymbol{B} = \boldsymbol{B}+\boldsymbol{A} \tag{2.10}$$

To add any number of vectors, the third is added to the sum of the first two, the fourth is then added to the resultant of this process and so on. Again, it is clear that

$$(\boldsymbol{A}+\boldsymbol{B})+\boldsymbol{C} = \boldsymbol{A}+(\boldsymbol{B}+\boldsymbol{C}) \tag{2.11}$$

so that vector addition follows the associative law.

So far we have discussed the manipulation of vectors without reference to any system of coordinates, but we may often wish to deal with the three resolved components of a vector parallel to the $x$-, $y$- and $z$-axes of a rectangular cartesian system. For this purpose we denote by $\boldsymbol{i}$, $\boldsymbol{j}$ and $\boldsymbol{k}$, vectors of unit magnitude which are parallel to the $x$-, $y$- and $z$-axes respectively.

If a vector $\boldsymbol{A}$ has magnitude $A$ and makes angles $\theta_x$, $\theta_y$ and $\theta_z$ with the three axes, the magnitudes of its resolved components will be

$$\begin{aligned} A_x &= A\cos\theta_x \\ A_y &= A\cos\theta_y \\ A_z &= A\cos\theta_z \end{aligned} \tag{2.12}$$

and the vectorial components themselves will be $A_x\boldsymbol{i}$, $A_y\boldsymbol{j}$ and $A_z\boldsymbol{k}$ respectively. Thus we may write

$$\boldsymbol{A} = A_x\boldsymbol{i}+A_y\boldsymbol{j}+A_z\boldsymbol{k} \tag{2.13}$$

These relations are illustrated in fig. 2.5.

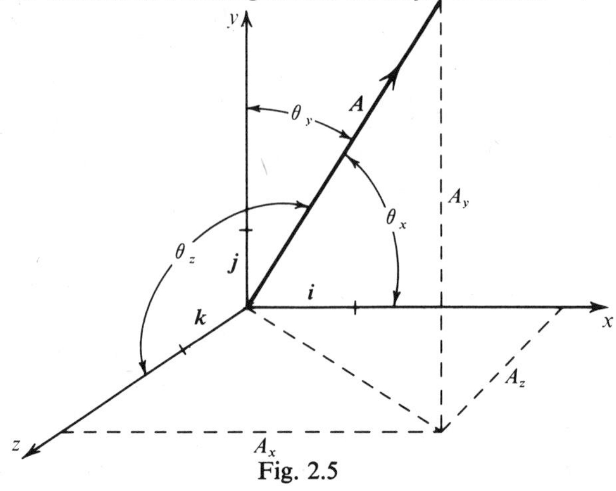


Fig. 2.5

We may wish to find the components of the sum of two vectors $\boldsymbol{A}$ and $\boldsymbol{B}$, when both are referred to the same set of axes. We have

$$\boldsymbol{A}+\boldsymbol{B} = (A_x\boldsymbol{i}+A_y\boldsymbol{j}+A_z\boldsymbol{k})+(B_x\boldsymbol{i}+B_y\boldsymbol{j}+B_z\boldsymbol{k})$$

or

$$\boldsymbol{A}+\boldsymbol{B} = (A_x+B_x)\boldsymbol{i}+(A_y+B_y)\boldsymbol{j}+(A_z+B_z)\boldsymbol{k} \tag{2.14}$$

Finally, it should perhaps be added that the vectors which we have been considering, and which represent quantities directed in space, are quite different from the vectors or phasors used in network theory to represent sinusoidally varying quantities which differ in phase.

### 2.3.2 The scalar product of two vectors

Suppose we have two vectors $\boldsymbol{A}$ and $\boldsymbol{B}$ whose directions are inclined at angle $\theta$. We are very often interested in the quantity $AB \cos \theta$; the product of the magnitude of the first vector and the resolved component, in the direction of the first vector, of the magnitude of the second. This quantity occurs so frequently that it is convenient to have a shorthand method of indicating it. We therefore define the *scalar* or *dot* product of two vectors by the identity

$$\boldsymbol{A}\cdot\boldsymbol{B} \equiv AB\cos\theta \tag{2.15}$$

The dot is an essential part of the notation and the product is termed scalar, because the result of the operation is a scalar quantity. We shall later meet a quite different vector product of two vectors.

When two vectors have the same direction, $\theta$ is zero and the scalar product is simply the numerical product of the magnitudes of the two vectors. In particular

$$\boldsymbol{A}\cdot\boldsymbol{A} = A^2 \tag{2.16}$$

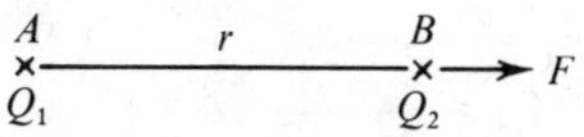

Fig. 2.6

Similarly, when two vectors are at right angles to each other, their scalar product is zero. Finally, it is clear from (2.15) that

$$\boldsymbol{A}\cdot\boldsymbol{B} = \boldsymbol{B}\cdot\boldsymbol{A} \tag{2.17}$$

If we are interested in the resolved components of the scalar product of two vectors, we have

$$\boldsymbol{A}\cdot\boldsymbol{B} = (A_x\boldsymbol{i}+A_y\boldsymbol{j}+A_z\boldsymbol{k})\cdot(B_x\boldsymbol{i}+B_y\boldsymbol{j}+B_z\boldsymbol{k})$$

However

$$\boldsymbol{i}\cdot\boldsymbol{i} = \boldsymbol{j}\cdot\boldsymbol{j} = \boldsymbol{k}\cdot\boldsymbol{k} = 1$$

and

$$\boldsymbol{i}\cdot\boldsymbol{j} = \boldsymbol{j}\cdot\boldsymbol{k} = \boldsymbol{k}\cdot\boldsymbol{i} = 0$$

so that

$$\boldsymbol{A}\cdot\boldsymbol{B} = A_xB_x+A_yB_y+A_zB_z \tag{2.18}$$

### 2.3.3 Other unit vectors

We have seen that it is convenient to specify directions along the three axes of a system of rectangular coordinates by means of the unit vectors $\boldsymbol{i}, \boldsymbol{j}$ and $\boldsymbol{k}$. We now introduce two other unit vectors, which we shall use later.

In fig. 2.6, suppose we have positive electric charges $Q_1$ and $Q_2$ situated at $A$ and $B$ respectively, where the length of $AB$ is equal to $r$. We shall see later that $Q_1$ exerts on $Q_2$ a repulsive force $F$, of magnitude $Q_1Q_2/kr^2$, where $k$ is a constant. In certain calculations we might wish to have a vector expression for $F$ which would indicate that it was acting along the direction of $r$. To this end we might write

$$\boldsymbol{F} = QQ_2\boldsymbol{r}/kr^3 \tag{2.19}$$

where we have multiplied the magnitude of the force by $\boldsymbol{r}/r$, which is a vector of unit magnitude. Alternatively, the method which we shall prefer is to write

$$\boldsymbol{F} = Q_1Q_2\boldsymbol{r}_0/kr^2 \tag{2.20}$$

where $\boldsymbol{r}_0$ is, by definition, the unit vector in the direction of $r$.

A second case arises when we wish to have a vector expression for both the area $S$ of a plane surface and the direction in which it lies. The direction is specified most readily by the direction of the normal to the surface, and we define a unit vector $\boldsymbol{n}$ in this direction. Our vector expression for the surface then becomes

$$\boldsymbol{S} = \boldsymbol{n}S \tag{2.21}$$

The application of this procedure will become apparent in the next section.

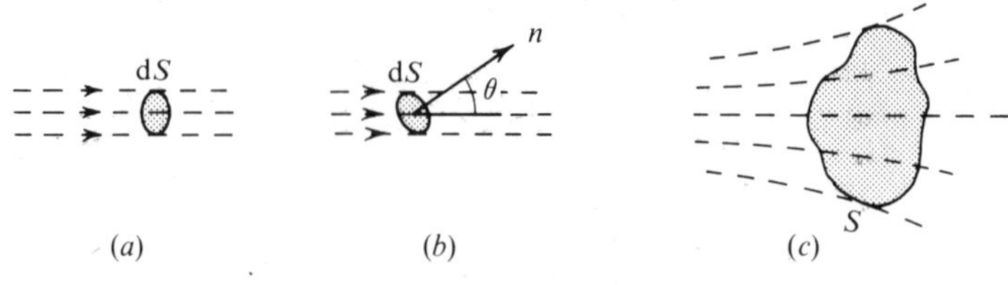


Fig. 2.7

## 2.4 The flow of current through a surface

### 2.4.1 Flow through an open surface

Suppose we have a medium in which current from any number of sources and sinks is flowing. We shall have complete knowledge of the distribution of current if we know the magnitude and direction of the current density $\boldsymbol{J}$ at every point. We now wish to derive a general expression for the current flowing through any given surface $S$, in terms of $\boldsymbol{J}$. In general, $S$ will not be plane, so we begin by considering an element $\mathrm{d}S$ which is sufficiently small to be considered plane. $\boldsymbol{J}$ is the current density, assumed constant over this small element.

In fig. 2.7($a$) we suppose the flow lines, shown dashed, to be normal to the element $\mathrm{d}S$. Then the current $\mathrm{d}I$ flowing through $\mathrm{d}S$ is given by

$$\mathrm{d}I = J\mathrm{d}S \tag{2.22}$$

Because we have already stipulated that $\mathrm{d}S$ lies at right angles to $\boldsymbol{J}$, there is no need to introduce vectors into our equation.

In the more general case, fig. 2.7($b$), $\mathrm{d}S$ will not be at right angles to the direction of flow. Let the normal to $\mathrm{d}S$ be $n$ and let it make angle $\theta$ with the flow direction. We now have

$$\mathrm{d}I = J\mathrm{d}S\cos\theta \tag{2.23}$$

which, in vector notation, becomes

$$\mathrm{d}I = \boldsymbol{J}\cdot\boldsymbol{n}\,\mathrm{d}S \tag{2.24}$$

where $\boldsymbol{n}$ is the unit vector in the direction of the normal.

Finally, in fig. 2.7($c$), we consider the flow through any surface $S$, which need no longer be small or plane. We divide this surface into small elements $\mathrm{d}S$ and find the flow through each by (2.24). The total current is then found by integration, to give

$$I = \int_S \boldsymbol{J}\cdot\boldsymbol{n}\,\mathrm{d}S \tag{2.25}$$

The integral sign, with $S$ at its base, is standard notation to indicate that the integration is to extend over the whole of the surface $S$. The right-hand

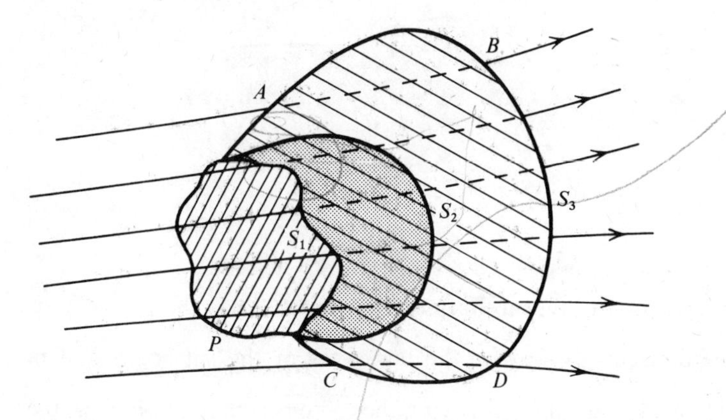


Fig. 2.8

side of (2.25) is termed a *surface integral.* It is worthy of note that although this integral involves two vectors, the left-hand side of (2.25) is a scalar quantity. This corresponds to the fact that the scalar product of two vectors is a scalar quantity. From a more physical point of view we note that, while $\boldsymbol{J}$ and $\boldsymbol{n}$ are vector quantities with definite directions at each point of the surface, $I$ is not; it is simply the total current flowing through the surface and has no unique direction.

Following a convention that we shall later use with other flux vectors, we may refer to the right-hand side of (2.24) as the flux of $\boldsymbol{J}$ passing through $\mathrm{d}S$. Similarly, the right-hand side of (2.25) is the flux of $\boldsymbol{J}$ passing through $S$.

## 2.4.2 A property of a flux vector

In fig. 2.8 let $S_1$ be a surface bounded by perimeter $P$, in a medium through which current is flowing. Let $I$ be the total current flowing through $S_1$. We now take a second bowl-shaped surface $S_2$, which has the same perimeter $P$ as $S_1$. It is clear that the current which passes through $S_2$ will be identical with that passing through $S_1$. Finally, consider a more bulbous surface $S_3$, which again has the same perimeter $P$. It is clear that all the current passing through $S_1$ will subsequently pass through $S_3$, but we must also take account of current represented by flow lines $AB$ or $CD$, which do not pass through $S_1$. However, all such current enters $S_3$ from one side and leaves to the same side, so it does not contribute to the total current through $S_3$. This is again equal to that through $S_1$. Thus we may write

$$I = \int_{S_1} \boldsymbol{J}\cdot\boldsymbol{n}\,\mathrm{d}S = \int_{S_2} \boldsymbol{J}\cdot\boldsymbol{n}\,\mathrm{d}S = \int_{S_3} \boldsymbol{J}\cdot\boldsymbol{n}\,\mathrm{d}S \tag{2.26}$$

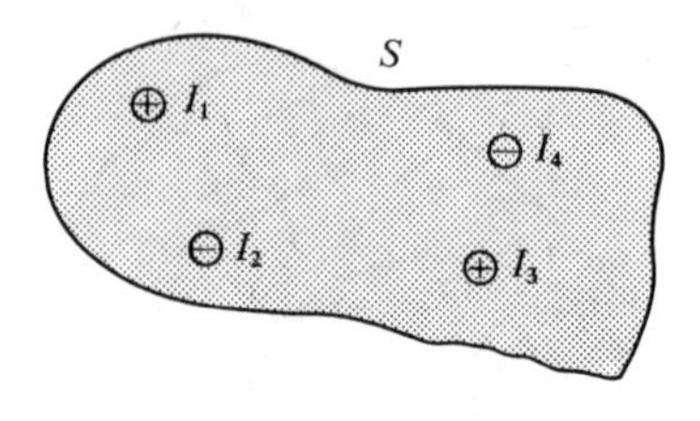


Fig. 2.9

We may now say that a flux vector such as $\boldsymbol{J}$ has the property that the $\int_S \boldsymbol{J}\cdot\boldsymbol{n}\,\mathrm{d}S$ depends only on the perimeter of the surface $S$ and has the same value for all surfaces having the same perimeter, so long as there are no sources or sinks in the relevant part of the medium. A vector which is not a flux vector does not possess this property and so it is meaningless to speak of the flux of such a vector through a surface which is defined only by its perimeter.

Conversely, any vector which possesses the above property is a flux vector.

### 2.4.3 Flow through a closed surface

Suppose we have a closed surface $S$, which contains sources of current (+) and sinks (−). In the steady state the algebraic sum of the currents entering and leaving the surface via the sources and sinks respectively must be equal to the total current passing through the surface. With the sources and sinks shown in fig. 2.9, we may write

$$\oint_S \boldsymbol{J}\cdot\boldsymbol{n}\,\mathrm{d}S = I_1+I_3-I_2-I_4 \tag{2.27}$$

where $\boldsymbol{n}$ must now be defined to be the outward-directed unit normal for each element of the surface. The circle on the integral sign indicates that the surface $S$ is closed. More generally we may write

$$\oint_S \boldsymbol{J}\cdot\boldsymbol{n}\,\mathrm{d}S = \Sigma I \tag{2.28}$$

In the special case where there are no sources or sinks within the surface

$$\oint_S \boldsymbol{J}\cdot\boldsymbol{n}\,\mathrm{d}S = 0 \tag{2.29}$$

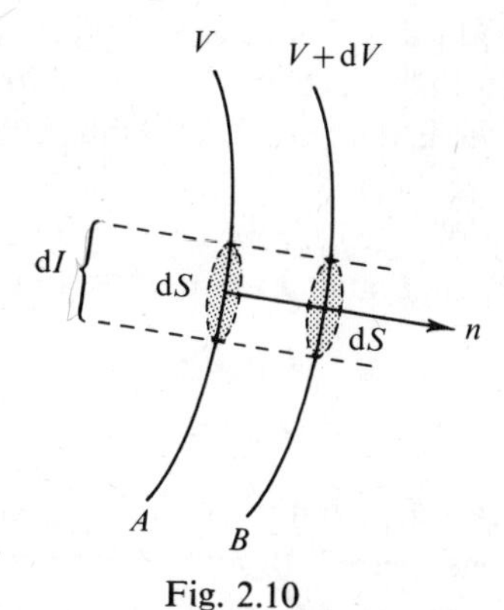


Fig. 2.10

## 2.5 The electric field strength

### 2.5.1 The definition of *E*

In fig. 2.10 let $A$ and $B$ be two adjacent equipotential surfaces in a medium through which current is flowing and let their respective potentials be $V$ and $V+\mathrm{d}V$. Since $\mathrm{d}V$ is to become negligibly small, we can draw a single normal $n$ to the two surfaces, in the direction in which $V$ is increasing. Let flow lines in the vicinity of this normal define a small volume of the medium through which current $\mathrm{d}I$ is flowing. This flow will be at right angles to the equipotentials, and hence parallel to $n$. However, current will flow from $B$ to $A$ since $B$ is at a higher potential than $A$. Thus the direction of $\mathrm{d}I$ is opposite to that which we have chosen as positive for $n$. Let the element of volume through which the current flows have length $\mathrm{d}n$ and area of cross-section $\mathrm{d}S$. Then from (2.2) its resistance will be $\rho\ \mathrm{d}n/\mathrm{d}S$ and we can write

$$\mathrm{d}I = -\frac{1}{\rho}\frac{\mathrm{d}V}{\mathrm{d}n}\,\mathrm{d}S$$

and current density is equal to

$$J = \frac{\mathrm{d}I}{\mathrm{d}S} = -\frac{1}{\rho}\frac{\mathrm{d}V}{\mathrm{d}n} \tag{2.30}$$

Since, in the region that we are considering, $\mathrm{d}n$ is the shortest distance between the two equipotentials, $\mathrm{d}V/\mathrm{d}n$ represents the maximum rate of change of $V$ with distance, in this region. Like $\boldsymbol{J}$ it is a vector quantity and has the same direction as the normal. To express these facts we write

$$\boldsymbol{J} = -\frac{1}{\rho}\frac{\mathrm{d}V}{\mathrm{d}n}\,\boldsymbol{n} \tag{2.31}$$

where $\boldsymbol{n}$ is the unit normal vector.

The quantity $(-\mathrm{d}V/\mathrm{d}n)\boldsymbol{n}$ is an important vector and it is convenient to give it a name. We shall term it the *electric field strength* and shall denote it by $\boldsymbol{E}$. It is also known as the electric field intensity. The negative sign on the right-hand sides of (2.30) and (2.31) expresses the fact that the direction of $\boldsymbol{J}$ is opposite to that of $(-\mathrm{d}V/\mathrm{d}n)\boldsymbol{n}$. The negative sign has been included in our definition of $\boldsymbol{E}$, so $\boldsymbol{E}$ and $\boldsymbol{J}$ are in the same direction and we may write

$$\boldsymbol{J} = \boldsymbol{E}/\rho \tag{2.32}$$

or

$$\boldsymbol{J} = \sigma\boldsymbol{E} \tag{2.33}$$

where $\sigma$, the reciprocal of $\rho$, is the *conductivity* of the medium. The SI unit for $\sigma$ is the siemens metre$^{-1}$ (S m$^{-1}$). We have already proved that $\boldsymbol{J}$ is a flux vector and, since $\rho$ and $\sigma$ are constants of the medium, it follows that $\boldsymbol{E}$ is also a flux vector in a single homogeneous isotropic medium.

In the next chapter, from a different point of view, we shall define $\boldsymbol{E}$ at any point to be the force acting on unit positive charge placed at that point. We have not used this definition in the present chapter because the macroscopic theory that we are developing is independent of any assumptions about the nature of the charged particles which carry the current. So long as Ohm's law and (2.2) are valid, our theory will apply. We are not concerned with forces on individual charged particles.

The SI unit for $\boldsymbol{E}$ is the *volt metre*$^{-1}$ (V m$^{-1}$).

### 2.5.2 $E$ and the gradient of $V$

We have already shown that the normal to an equipotential is the direction in which $V$ is changing most rapidly with distance. In consequence $(\mathrm{d}V/\mathrm{d}n)\boldsymbol{n}$ is often termed the *gradient* of $V$. Thus, in vector notation, we write

$$\boldsymbol{E} = -\frac{\mathrm{d}V}{\mathrm{d}n}\boldsymbol{n} = -\mathrm{grad}\ V \tag{2.34}$$

When we work with rectangular coordinates, we may wish to know the components of $\boldsymbol{E}$ along the directions of the three axes. We then have

$$E_x = -\frac{\partial V}{\partial x}\boldsymbol{i},\quad E_y = -\frac{\partial V}{\partial y}\boldsymbol{j},\quad E_z = -\frac{\partial V}{\partial z}\boldsymbol{k} \tag{2.35}$$

These three components, when added vectorially, give the total field strength $\boldsymbol{E}$. Hence, from (2.34) and (2.35),

$$\mathrm{grad}\ V = \boldsymbol{i}\frac{\partial V}{\partial x} + \boldsymbol{j}\frac{\partial V}{\partial y} + \boldsymbol{k}\frac{\partial V}{\partial z} \tag{2.36}$$

## 2.6 Problems

1. In fig. 2.2, prove that the equipotential lines must intersect the edges of the conducting sheet at right angles.

2. Two vectors $\boldsymbol{A}$ and $\boldsymbol{B}$ are such that

$$\boldsymbol{A}+\boldsymbol{B} = 3\boldsymbol{i}+4\boldsymbol{j}-7\boldsymbol{k}$$

$$\boldsymbol{A}-\boldsymbol{B} = 5\boldsymbol{i}-2\boldsymbol{j}+3\boldsymbol{k}$$

Find the components of $\boldsymbol{A}$ and $\boldsymbol{B}$ and the magnitudes of these two vectors. Find also the values of $\boldsymbol{A}\cdot\boldsymbol{B}$ and the angle between $\boldsymbol{A}$ and $\boldsymbol{B}$.

3. In a concentric cable the external radius of the inner conductor is 1 mm and the internal radius of the outer conductor is 1 cm. The space between the conductors is filled with insulating material of resistivity $10^{13}$ Ω m. If the voltage between the conductors is 100 kV, what will be the current flow per kilometre of cable?

4. A metal hemisphere of radius $R$, buried with its flat face lying in the surface of the ground, is used as an earthing electrode. It may be assumed that a current flowing to earth spreads out uniformly and radially from the electrode for a great distance. Show that, as the distance for which this is true tends to infinity, the resistance between the electrode and earth tends to the limiting value $\rho/2\pi R$, where $\rho$ is the resistivity of the earth.

If $R = 0.5$ m and $\rho = 100$ Ω m, what will be the resistance between two such electrodes, situated a great distance apart?

If a fault current of 1000 A flows to earth through an electrode of this kind, what will be the maximum potential difference between two points on the earth's surface, 0.5 m apart, if the mean distance of the points from the electrode is (*a*) 100 m, (*b*) 1000 m?

# 3

# The electrostatic field in free space

## 3.1 Introduction

### 3.1.1 The basic experimental law

In this chapter we shall deal with the interaction of stationary electric charges in free space. Experiment shows that a force exists between any two bodies which carry nett electric charges (i.e. an excess of positive over negative charge or *vice versa*) when these bodies are at rest. Such forces are termed *electrostatic*. We shall see later that additional forces arise from charges in motion.

Electrostatic forces were studied by Coulomb who, as a result of his experiments, enunciated the following law.

If two charges of magnitude $Q_1$ and $Q_2$ respectively are situated on bodies whose dimensions are small compared with the distance $d$ between them, the resulting electrostatic force

(*a*) is proportional to $Q_1Q_2$,
(*b*) is inversely proportional to $d^2$,
(*c*) acts along the line joining the bodies,
(*d*) is attractive or repulsive according as $Q_1$ and $Q_2$ are of unlike or like sign.

Thus
$$\text{Force} = F = kQ_1Q_2/d^2 \tag{3.1}$$
where $k$ is a constant.

Coulomb measured forces by means of a torsion balance and his experiments were crude by modern standards. More recently, much more accurate methods of verifying (3.1) have been devised (§10.1.2) and it has been shown that, if any error in the relation does exist, it cannot be greater than one part in $10^9$.

### 3.1.2 Units

The SI unit of length is the metre and we have already defined the unit of charge, the coulomb (§2.1.1). The unit of force is the *newton* (symbol N) and the constant $k$ in (3.1) must be consistent with these units.

We shall now re-write (3.1) in the form
$$F = Q_1Q_2/4\pi\epsilon_0 d^2 \tag{3.2}$$

The rather arbitrary insertion of $4\pi$ in the denominator has the advantage that future equations relating to systems with spherical symmetry will usually contain the constant $4\pi$; when the systems have cylindrical symmetry the constant will be $2\pi$, and when the symmetry is planar, the constant will be unity. This is an aid to memory, but the advantage is perhaps not overwhelming. What is important is that this so-called *rationalized system* of units has been adopted by international agreement.

The constant $\epsilon_0$ in (3.2) is known as the *permittivity of free space*, the *permittivity of a vacuum*, or the *electric constant*. We shall use the first of these names. Its value must be determined by experiment.

If we attempt to derive the SI unit for permittivity from (3.2), we arrive at the rather cumbersome (coulomb)$^2$ (second)$^2$ (kilogram)$^{-1}$(metre)$^{-3}$. We shall later (§3.7.1) define a unit of capacitance, the *farad* (symbol F). From this definition the reader may verify that an equivalent SI unit for permittivity is the *farad metre*$^{-1}$ (symbol F m$^{-1}$) and this unit is commonly used.

### 3.1.3 The value of $\epsilon_0$

In principle it is possible to determine the value of $\epsilon_0$ by direct measurement of the force between two charged bodies, but a much more accurate result can be obtained by indirect means. According to Maxwell's electromagnetic theory of light (chapter 9), the velocity of light in a vacuum, $c$, is related to $\epsilon_0$ by the equation

$$\mu_0 \epsilon_0 = 1/c^2 \tag{3.3}$$

where $\mu_0$ is another constant which we shall term the *permeability of free space*. It is also known as the *permeability of a vacuum* or the *magnetic constant*. Furthermore, we shall show later (§4.3.2) that $\mu_0$ is involved in the definition of the ampere (§2.1.1) and that, by that definition, $\mu_0$ must have the *exact* value

$$\mu_0 = 4\pi \times 10^{-7}\ \text{H m}^{-1} \tag{3.4}$$

Combining this result with (3.3) and taking for the velocity of light the experimental value

$$c = 2.997925 \times 10^{8}\ \text{m s}^{-1} \tag{3.5}$$

we have

$$\epsilon_0 = 8.854185 \times 10^{-12}\ \text{F m}^{-1} \tag{3.6}$$

For many purposes it is sufficiently accurate to take $c$ equal to $3 \times 10^8$ m s$^{-1}$ and we then have *approximately*

$$\epsilon_0 = 10^{-9}/36\pi\ \text{F m}^{-1} \tag{3.7}$$

The unit for $\mu_0$ will be explained in §4.1.5.

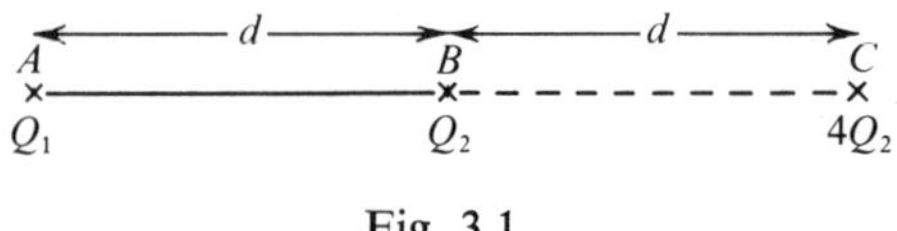


Fig. 3.1

### 3.1.4 The principle of superposition

When more than two charges are present in free space, any one of them will experience a force which is the vector sum of the forces resulting from each of the others, taken in turn. This is the *principle of superposition*. Its truth rests on experimental evidence.

If we are dealing with charged bodies whose linear dimensions are not small compared with the distances separating them, we must consider each body to carry a number of point charges and then use the principle of superposition to add the forces resulting from the charges taken separately.

## 3.2 The electric field

### 3.2.1 Definition of $E$

In fig. 3.1 let charges $Q_1$ and $Q_2$ be placed on small bodies at $A$ and $B$ respectively, where the distance between $A$ and $B$ is $d$. Then $Q_1$ experiences an electrostatic force (3.2)

$$F = Q_1 Q_2 / 4\pi\epsilon_0 d^2 \tag{3.8}$$

If we knew only the magnitude and direction of this force, we could not assert that it had been caused by charge $Q_2$ at distance $d$. It might equally well have been caused by a charge $4Q_2$ at distance $2d$, or by any other charge at the appropriate distance. Again, it might have arisen from not one but any number of charges distributed throughout space in an infinite variety of possible ways. We may express this fact by saying that each of these distributions produces the same value of the *electric field strength* $\boldsymbol{E}$ at point $A$, and that the force on $Q_1$ is produced by the interaction of $Q_1$ with the electric field. From this point of view we define $\boldsymbol{E}$ at any point of an electric field to be the force acting on unit positive charge placed at that point. We must make the proviso that the introduction of the unit charge to measure $\boldsymbol{E}$ must not in any way disturb the distribution of the existing charges in the field. Thus, more accurately, we should write

$$\boldsymbol{E} = \mathrm{Lt}_{Q \to 0} \boldsymbol{F}/Q \tag{3.9}$$

For reasons which we shall shortly discuss (§3.3.1), the SI unit of $\boldsymbol{E}$ is the volt per metre ($\mathrm{V\ m^{-1}}$).

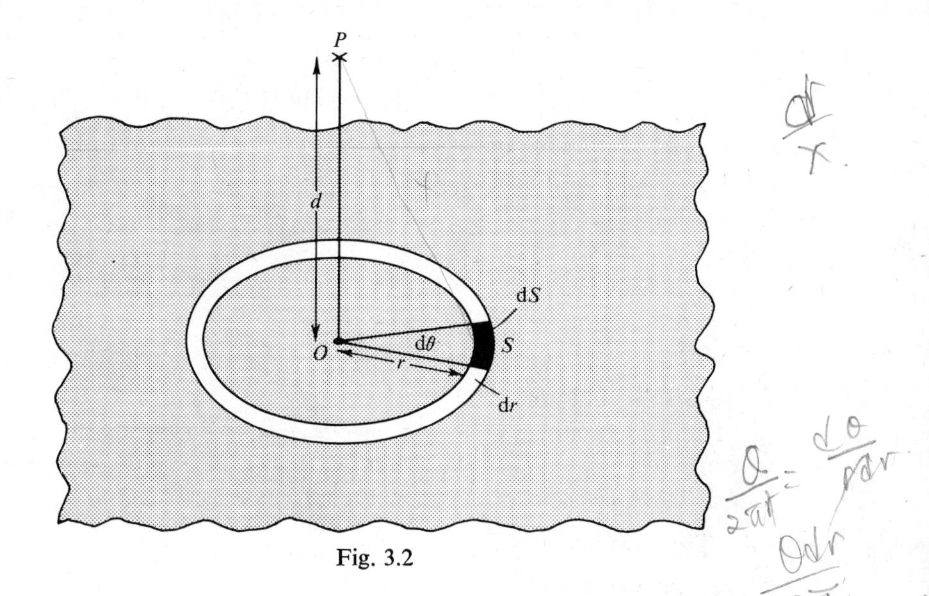


Fig. 3.2

### 3.2.2 The electric field caused by an infinite plane sheet of charge

As an example of the calculation of the electric field caused by a continuous distribution of charge, we find the value of $\boldsymbol{E}$ when charge is distributed uniformly over an infinite plane conductor.

In fig. 3.2 let $P$ be a point at distance $d$ above the plane conductor and let the density of charge be $Q$ coulombs per square metre. Let $O$ be the foot of the perpendicular from $P$ on to the plane. Consider the charge lying between circles of radii $r$ and $r+\mathrm{d}r$ respectively, with $O$ as centre. The area of the element $\mathrm{d}S$ at $S$ of the surface bounded by these two circles and by two radii which make angle $\mathrm{d}\theta$ with each other is $r\,\mathrm{d}\theta\,\mathrm{d}r$, and the charge on it is $Qr\,\mathrm{d}\theta\,\mathrm{d}r$. This charge will exert a force on unit charge placed at $P$, of magnitude

$$Qr\,\mathrm{d}\theta\,\mathrm{d}r/4\pi\epsilon_0(d^2+r^2)$$

and directed along $SP$. The force will have a component at right angles to the plane, of magnitude

$$\frac{Qr\,\mathrm{d}\theta\,\mathrm{d}r}{4\pi\epsilon_0(d^2+r^2)}\,\frac{d}{(d^2+r^2)^{\frac{1}{2}}} = \frac{Qrd\,\mathrm{d}\theta\,\mathrm{d}r}{4\pi\epsilon_0(d^2+r^2)^{\frac{3}{2}}}$$

It will also have a component parallel to the plane but, when we consider the complete ring of charge, it is clear that these components will cancel. Thus putting $\mathrm{d}\theta$ equal to $2\pi$ for the complete ring of charge, the force normal to the plane on unit charge at $P$ is

$$\frac{2\pi Qr d\mathrm{d}r}{4\pi\epsilon_0(d^2+r^2)^{\frac{3}{2}}} = \frac{Qd}{2\epsilon_0}\frac{r\,\mathrm{d}r}{(d^2+r^2)^{\frac{3}{2}}}$$

Finally, integrating for the whole plane,

$$E = \frac{Qd}{2\epsilon_0}\int_0^\infty \frac{r\,\mathrm{d}r}{(d^2+r^2)^{\frac{3}{2}}} = \frac{Q}{2\epsilon_0} \tag{3.10}$$

Since $E$ is directed at right angles to the plane, we write

$$\boldsymbol{E} = \frac{Q}{2\epsilon_0}\boldsymbol{n} \tag{3.11}$$

where $\boldsymbol{n}$ is the unit normal.

It is worthy of note that, in this case, $\boldsymbol{E}$ is independent of $d$. In practice we cannot have an infinite plane, but (3.11) should be very nearly true so long as the dimensions of the plane are large compared with $d$.

### 3.2.3 Lines of $E$

Since $\boldsymbol{E}$ is the force on unit charge, it has a unique direction at each point of the field, unless it is zero. It is thus possible to draw lines to represent the direction of $\boldsymbol{E}$, similar to the flow lines which we used in the preceding chapter. However, it is not obvious that these lines can be drawn in such a way as to represent also the magnitude of $\boldsymbol{E}$, since we have not yet proved that $\boldsymbol{E}$ is a flux vector. Moreover, the proof that was used in the case of the conducting medium will not serve in the present instance since, in the electrostatic field, there is no tangible quantity, such as charge, which flows through the field without loss. Any deduction that we make about the vector $\boldsymbol{E}$ must be based on the inverse square law (3.1). The proof that we need is provided by a theorem due to Gauss, but we first digress to explain the notation of solid angles.

### 3.2.4 Solid angles

Just as, in plane geometry, we deal with angles so, in solid geometry, we have need of a quantitative measure of a solid angle. The latter may be defined as the space enclosed by a conical surface generated by straight lines through a point, though not necessarily of circular cross-section. The simplest way to arrive at a measure for a solid angle is by analogy with the planar case.

In fig. 3.3(*a*), if we wish to find the size of angle $AOB$ in radian measure,

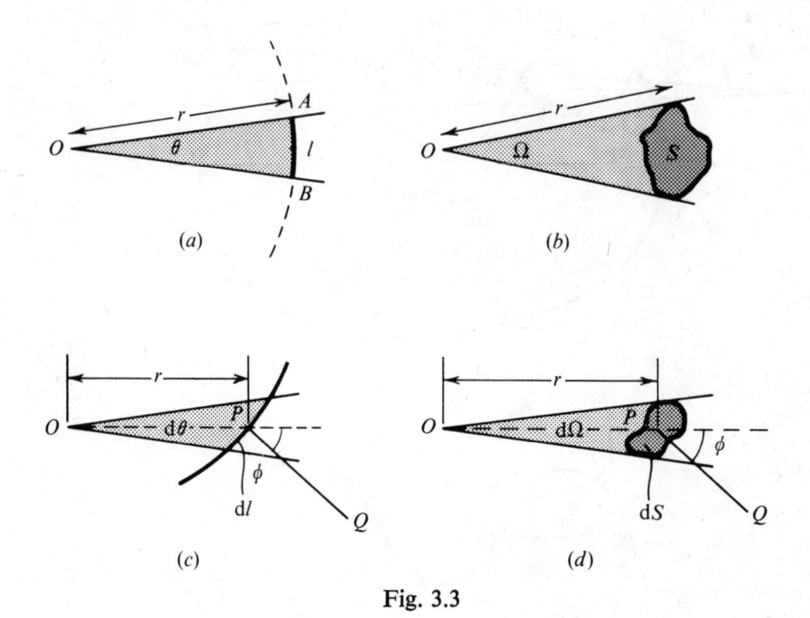


Fig. 3.3

we describe a circle of radius $r$ with centre $O$, from which the angle cuts off an arc of length $l$. We then say that the size of the angle $AOB$ is

$$\theta = l/r \text{ radians} \tag{3.12}$$

Similarly, in fig. 3.3(*b*), to measure the size of a solid angle $\Omega$, we describe a sphere of radius $r$, whose centre is the apex $O$ of the cone containing $\Omega$. From this sphere the cone cuts out an area $S$ and the size of the solid angle $\Omega$ in *steradians* (symbol sr) is given by

$$\Omega = S/r^2 \tag{3.13}$$

Just as, in the planar case, the total angle contained by a complete circle is $2\pi$ rad so, in solid geometry, the total solid angle filling a complete sphere is

$$\text{area of surface of sphere}/r^2 = 4\pi \text{ sr} \tag{3.14}$$

Further useful expressions can be obtained when the angles under consideration are infinitesimally small. In fig. 3.3(*c*) let the angle $d\theta$ intersect a curve of any shape, cutting out at $P$ a length $dl$ which, in the limit, will be straight, and let $r$ be the distance of $P$ from $O$. Let $PQ$ be the normal to $dl$, making angle $\phi$ with the continuation of $OP$. Then

$$d\theta = (dl \cos \phi)/r \text{ rad} \tag{3.15}$$

Similarly, in the solid case, fig. 3.3(*d*). the infinitesimal solid angle $d\Omega$ cuts

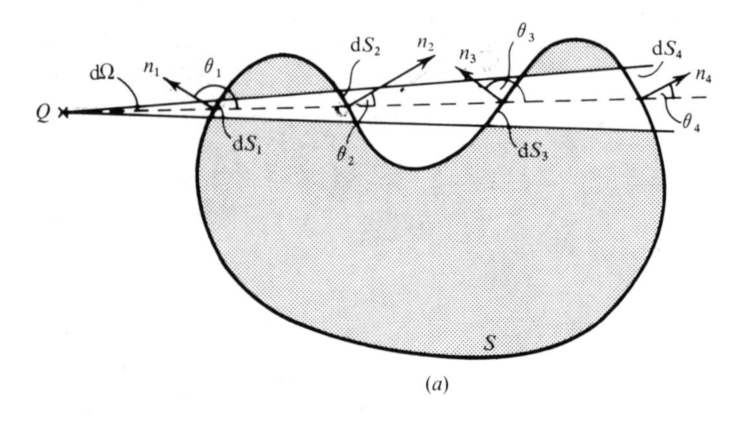


(*a*)

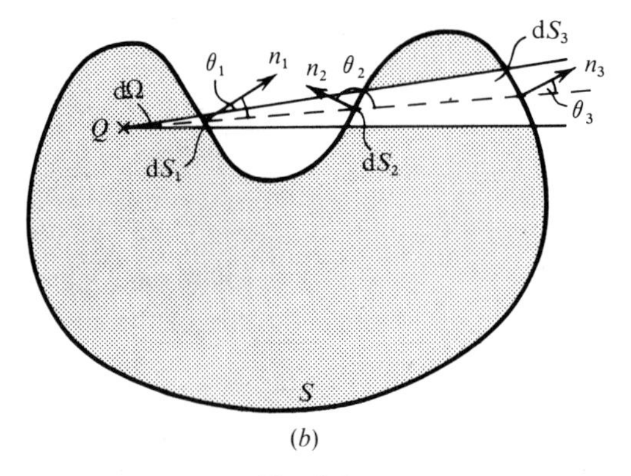


(*b*)

Fig. 3.4

out an area d$S$ from a curved surface of any shape. In the limit d$S$ will be plane; let the normal to it be $PQ$, making angle $\phi$ with the axis of the cone. Then

$$d\Omega = (dS \cos \phi)/r^2 \text{ sr} \tag{3.16}$$

### 3.2.5 Gauss' theorem

In fig. 3.4(*a*) let a point charge $Q$ be situated outside a closed surface $S$. From $Q$ let an infinitesimally small cone, of solid angle d$\Omega$, cut out areas d$S_1$, d$S_2$, d$S_3$ and d$S_4$ from $S$ and let $n_1$, $n_2$, $n_3$ and $n_4$ be the normals to these areas, making angles $\theta_1$, $\theta_2$, $\theta_3$ and $\theta_4$ with the axis of the cone. Let the areas be distant $r_1$, $r_2$, $r_3$ and $r_4$ respectively from $Q$.

$Q$ will give rise to an electric field of strength $\boldsymbol{E}$ and, at each of our small elements of area, this field will be directed along the axis of the cone. We wish to calculate the outward-going flux of $\boldsymbol{E}$ through each of the areas.

It is to be noted that, while it is proper to speak of the flux of $\boldsymbol{E}$ through an element of area at a particular point, we must not make any assumption about the properties of $\boldsymbol{E}$ over an extended region of space; we have not yet proved that $\boldsymbol{E}$ is a flux vector.

The value of the electric field strength at $\mathrm{d}S_1$ is

$$E_1 = Q/4\pi\epsilon_0 r_1^2 \tag{3.17}$$

and the flux of $\boldsymbol{E}$ through $\mathrm{d}S_1$ is therefore (cf §2.4.1)

$$E_1 \mathrm{d}S \cos\theta_1 = Q\,\mathrm{d}S\cos\theta/4\pi\epsilon_0 r_1^2 = -Q\,\mathrm{d}\Omega/4\pi\epsilon_0 \tag{3.18}$$

The negative sign in the last term arises from the fact that, with our convention as to positive directions (outward-drawn normal to the surface and outward-going axis of the cone from $O$), $\theta$ is greater than $\pi/2$. It corresponds to the fact that the flux of $\boldsymbol{E}$ passing outward through $\mathrm{d}S_1$ is clearly negative.

Similar calculations can be made for the other elemental surfaces $\mathrm{d}S_2$, $\mathrm{d}S_3$ and $\mathrm{d}S_4$. For $\mathrm{d}S_2$ and $\mathrm{d}S_4$ the outward flux of $\boldsymbol{E}$ will be $Q\,\mathrm{d}\Omega/4\pi\epsilon_0$ while for $\mathrm{d}S_3$, as for $\mathrm{d}S_1$, it is $-Q\,\mathrm{d}\Omega/4\pi\epsilon_0$. Thus, for the sum for all four elements, the total outward flux of $\boldsymbol{E}$ is zero. Taking a sufficient number of elemental solid angles we can cover the whole of the surface $S$ and we therefore conclude that a point charge situated outside $S$ contributes nothing to the total outward flux of $\boldsymbol{E}$ through $S$.

Consider next, fig. 3.4(*b*), what happens when the charge $Q$ is inside $S$. We take elemental cones and proceed as before finding for $\mathrm{d}S_1$ and $\mathrm{d}S_3$ the outward flux to be $Q\,\mathrm{d}\Omega/4\pi\epsilon_0$, while for $\mathrm{d}S_2$ it is $-Q\,\mathrm{d}\Omega/4\pi\epsilon_0$. Hence the total contribution to the flux from areas intersected by the cone is $Q\,\mathrm{d}\Omega/4\pi\epsilon_0$. For the whole surface, the total flux will be

$$\frac{Q}{4\pi\epsilon_0}\oint_S \mathrm{d}\Omega = \frac{Q}{\epsilon_0} \tag{3.19}$$

because the whole closed surface will subtend a complete solid angle of $4\pi$ steradians (3.14). Using the expression developed in (§2.4.1) for flux through a surface, we may write

$$\oint_S \boldsymbol{E}\cdot\boldsymbol{n}\,\mathrm{d}S = Q/\epsilon_0 \tag{3.20}$$

So far we have proved this result only for a single point charge $Q$ but, by the superposition principle (§3.1.4), it will hold for any number of charges. At any element of surface it is only the normal component of $\boldsymbol{E}$ which causes flux to pass through the surface and these components will add algebraically. Continuous distributions of charge can be considered to be

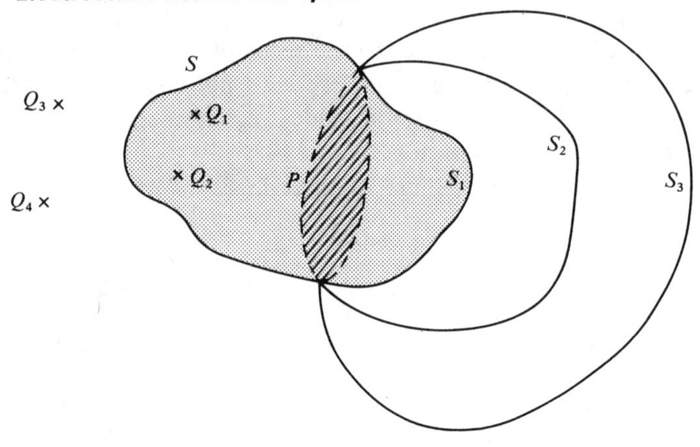


Fig. 3.5

assemblies of point charges so, for all cases, we may write

$$\oint_S \boldsymbol{E}\cdot\boldsymbol{n}\mathrm{d}S = \frac{1}{\epsilon_0}\Sigma Q \tag{3.21}$$

where $\Sigma Q$ is the total charge inside $S$. This is Gauss' theorem, which may be stated as *the total outward flux of* $\boldsymbol{E}$ *over any closed surface is equal to the algebraic sum of all charges within the surface, divided by* $\epsilon_0$.

If we construct a closed surface immediately outside a charge $Q$ the flux through this surface will be $Q/\epsilon_0$ and will be directed outward or inward according as $Q$ is positive or negative. Thus we may say that flux $1/\epsilon_0$ begins on each coulomb of positive charge and ends on each coulomb of negative charge.

### 3.2.6 $E$ as a flux vector

It follows from (3.21) that $\boldsymbol{E}$ is a flux vector in any region where there is no charge. In fig. 3.5, let $S$ be a closed surface which contains some charges, such as $Q_1$, $Q_2$, but not others, such as $Q_3$, $Q_4$. On $S$ draw any closed line $P$ (shown dashed) and let the portion of $S$ to the right of $P$ be termed $S_1$. The total outward flux through $S$ is, by (3.21), determined solely by the charges inside and would not be altered if $S_1$ were replaced by some other surface, such as $S_2$ or $S_3$, which had the same perimeter $P$ as $S_1$, so long as the new surface did not include any new charge. Thus, with this proviso, the flux of $\boldsymbol{E}$ through any surface bounded by $P$ is independent of the nature of that surface. By suitable choice of $S$ we can give $P$ any form and place it in any part of the field where there is no charge, without invalidating our result. This, however, can only be true if $\boldsymbol{E}$ is a flux vector (cf. §2.4.2).

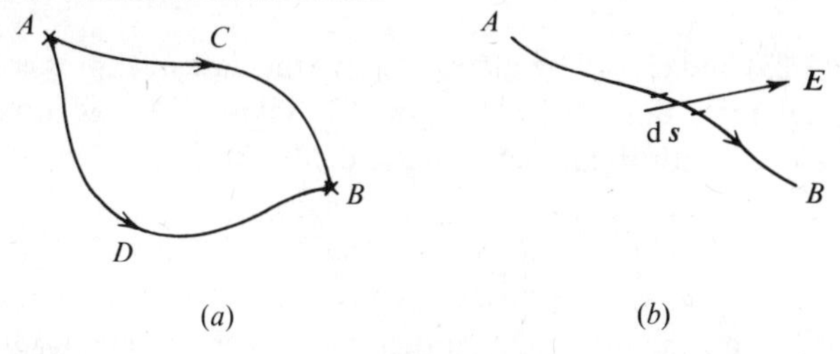


Fig. 3.6

## 3.3 Potential in the electrostatic field

### 3.3.1 The definition of potential difference

In fig. 3.6(*a*) let $A$ and $B$ be any two points in an electrostatic field and consider the work done in conveying unit positive charge from $A$ to $B$. If possible, let the work be greater when the charge is taken along the path $ACB$ than when it goes along $ADB$. Then, if we take the charge round the complete circuit $ADBCA$, the work given out on the return journey will be greater than that absorbed during the forward journey, and there is a nett gain of energy although the charge is in its original state. This, however, is impossible, so we conclude that the work needed to take unit charge from $A$ to $B$ is quite independent of the path travelled. It is therefore legitimate to refer to this work as the *potential difference* between $A$ and $B$, without specifying any particular path. Furthermore, if we make some arbitrary choice of point at which the potential is to be taken to be zero, we can attach a unique value of potential to every other point of the field.

In fig. 3.6(*b*) let $\mathrm{d}\boldsymbol{s}$ be an element of one particular path between $A$ and $B$ and let $\boldsymbol{E}$ be the electric field strength at this element. Then the work given out when unit positive charge is taken along $\mathrm{d}\boldsymbol{s}$ in the direction from $A$ to $B$ is $\boldsymbol{E}\cdot\mathrm{d}\boldsymbol{s}$. Since the potential $V_A$ at $A$ exceeds the value $V_B$ at $B$ by the work given out in taking unit positive charge from $A$ to $B$,

$$V_A - V_B = \int_A^B \boldsymbol{E}\cdot\mathrm{d}\boldsymbol{s} \tag{3.22}$$

We can construct equipotential surfaces in the electrostatic field in exactly the same way as in a conducting medium through which current is flowing (§2.1.3) and, using the same arguments as before (§2.5.1 and §2.5.2), we can show that

$$\boldsymbol{E} = -\frac{\mathrm{d}V}{\mathrm{d}n}\boldsymbol{n} = -\operatorname{grad} V \tag{3.23}$$

or, in rectangular coordinates,

$$E_x = -\frac{\partial V}{\partial x}\boldsymbol{i}, \quad E_y = -\frac{\partial V}{\partial y}\boldsymbol{j}, \quad E_z = -\frac{\partial V}{\partial z}\boldsymbol{k} \tag{3.24}$$

Equations (3.23) and (3.24) are alternative mathematical expressions of the same physical fact, that the work done in taking unit positive charge round any closed path in an electrostatic field is zero, or

$$\oint_S \boldsymbol{E}\cdot d\boldsymbol{s} = 0 \tag{3.25}$$

where the circle on the integral sign denotes that the integration is round a closed path. Integrals such as those on the right-hand side of (3.22) and the left-hand side of (3.25) are known as *line integrals*.

The physical fact of which (3.25) is an expression is a consequence of the conservation of energy. Thus the vector field of a quantity like $\boldsymbol{E}$, which satisfies an equation such as (3.22), is said to be a *conservative field*. A field of this kind has the property that, at every point, the magnitude and direction of the vector can be expressed as the derivative of a scalar quantity which has a unique value at that point. We shall see later that some vector fields do not possess this property.

### 3.3.2 The zero of potential

So far, we have defined only differences of potential; absolute potential can be defined only if we choose some reference point from which to measure the work done. The choice is quite arbitrary and, in any practical problem, unimportant, since we are invariably concerned only with potential differences.

For theoretical calculations it is often convenient to take as our zero of potential that of a point at an infinite distance from all charges, where $V$ might be expected to be zero. The potential of any other point is then the work done in bringing unit positive charge from infinity to that point. Occasionally this choice of zero leads to an apparent paradox, in that the potentials at certain points of the field become infinite. This occurs, for example, when we attempt to calculate the potential of some idealized systems, such as a uniformly charged rod of infinite length. It has not the slightest physical significance and the difficulty can always be avoided by choosing a more appropriate point for the zero of potential. This will not affect potential differences which are all that we really wish to know.

In practical problems it is sometimes convenient to choose the earth as our zero of potential. Reasons for doing this are that the earth is a conductor, and an experimenter, standing upon the earth, is likely to be approximately at earth potential. The earth is the largest conductor with which we have to deal and, in general, we cannot get far away from it. Apart from these purely practical considerations, which may or may not be important, the earth has no special significance; it is simply a conductor in the vicinity of most systems that we have to consider.

In this connection, a word of warning should perhaps be given. The earth is not a very good conductor and, when heavy currents are flowing through it, considerable differences of potential can occur between one point and another. This can happen during lightning strokes or when insulation fails in power systems. Engineers dealing with transatlantic telephone cables have found that during severe storms a potential difference as great as 1000 V can exist between Great Britain and the United States!

### 3.3.3 Comparison of electric fields

It will be clear that there is great similarity between the electric field that exists in a conducting medium through which current is flowing and the electrostatic field resulting from a distribution of charges at rest. In both cases we have been able to define a scalar potential $V$ and a vector electric field strength $\boldsymbol{E}$ and these two quantities have the same properties and relations in the two cases. Nevertheless, the paths by which we arrived at the definitions of these quantities are not at all the same.

In the case of the conducting medium we took Ohm's law and the indestructibility of current (except at sources or sinks) as our basic facts. From these we were able to show that $\boldsymbol{J}$ is a flux vector and to discuss the properties of such vectors and their relationships with equipotential surfaces. $\boldsymbol{E}$ was then defined as the derivative of the potential and, because of its constant proportionality to $\boldsymbol{J}$, was found to be a flux vector also.

In the electrostatic field, we began with the experimental inverse-square law of Coulomb, which led immediately to a definition of $\boldsymbol{E}$. In this case there is no physical entity such as current, which is being conveyed without loss through the field. Nevertheless, we were able to show, by way of Gauss' theorem, that $\boldsymbol{E}$ has the properties of a flux vector. Finally, we defined potential difference between two points as the line integral of $\boldsymbol{E}$ along a path joining the points.

## 3.4 Electric displacement

In our description of conductivity in an extended medium we found it convenient to introduce, in addition to the scalar potential $V$, two vector quantities; the current density $\boldsymbol{J}$ and the electric field strength $\boldsymbol{E}$, related by the equation

$$\boldsymbol{J} = \sigma \boldsymbol{E} \tag{3.26}$$

where $\sigma$ is the conductivity of the medium.

So far in our description of the electrostatic field we have needed only one vector quantity, the field strength $\boldsymbol{E}$, and, if our investigations were confined to fields in free space, this single quantity would suffice. However, when we come to consider electrostatic fields in material media we shall find it convenient to make use of a second vector, and we therefore introduce this vector into the discussion of fields in free space also.

In free space we define the *electric displacement* vector $\boldsymbol{D}$ by the relation

$$\boldsymbol{D} = \epsilon_0 \boldsymbol{E} \tag{3.27}$$

and the analogy with (3.26) is obvious. Since $\epsilon_0$ is a scalar constant, $\boldsymbol{E}$ and $\boldsymbol{D}$ are always in the same direction and $\boldsymbol{D}$, like $\boldsymbol{E}$, is a flux vector.

From (3.21), we have for any closed surface $S$,

$$\oint_S \boldsymbol{D}\cdot\boldsymbol{n}\,\mathrm{d}S = \epsilon_0 \oint_S \boldsymbol{E}\cdot\boldsymbol{n}\,\mathrm{d}S = \Sigma Q \tag{3.28}$$

i.e., the total outward flux of $\boldsymbol{D}$ through any closed surface is equal to the algebraic sum of the enclosed charges. By taking a surface immediately outside any particular charge, we may conclude that unit flux of $\boldsymbol{D}$ begins on a unit positive charge and ends on a unit negative charge.

We see from (3.28) that $\boldsymbol{D}$ has the dimensions of charge per unit area: the unit in which it is measured is therefore the *coulomb metre*$^{-2}$ (C m$^{-2}$).

## 3.5 The nature of electrostatic field problems

If we know the magnitude and direction of $\boldsymbol{E}$ at every point of an electrostatic field, we have complete information about that field. Thus the determination of $\boldsymbol{E}$ must be one of our basic aims. We have seen that, in principle, this determination can be carried out if the position and magnitude of each charge in the field is known, though the calculation may be very laborious. In the earlier part of this chapter we have often tacitly assumed that complete knowledge of the charges is available but, in practical problems, this is rarely the case for reasons which we now consider.

In general, the charges with which we have to deal reside on material bodies which may be conductors or insulators and the situation is different in the two cases. There are several ways in which bodies may become charged (e.g. by friction, by connection to a battery, or by the collection of free electrons), but, as a rule, there is no simple experimental method by which a measured quantity of charge can be conveyed to a body. Thus our knowledge of the magnitudes of the charges in the field is likely to be unsatisfactory. Even if this difficulty could be overcome, we should be faced with other problems.

If the body on which a charge resides is an insulator, we shall generally have no means of knowing how the charge is distributed over the surface or throughout the volume of the body. The problem of calculating the effect of the charge on the electrostatic field is then insoluble and we shall not consider this case any further. This does not mean that charges on insulators are unimportant; their practical effects can be serious and it is often necessary to take steps to prevent insulators from becoming charged.

If the charged body is a conductor, we shall again generally be ignorant of the distribution of the charge over the body. We certainly cannot assume that it will spread uniformly over the surface. What it must do is to distribute itself in such a way as to bring all parts of the conductor to the same potential. Under static conditions, a difference of potential cannot exist in a conductor; any initial differences which are present when the body is charged will cause currents to flow until uniformity of potential has been established.

From the above discussion it should be clear that, if we wish to set up an electrostatic field of known properties, the charged bodies should be conductors. Furthermore, although we cannot readily convey measured quantities of charge to these conductors we can, by connecting them to batteries, maintain them at accurately known potentials with respect to each other. A field set up in this way is repeatable at any time. Thus our ultimate problem must be to determine the properties of a field which results from the application of known potentials to conductors of known sizes, shapes and positions. The general solution of this problem will be considered in chapter 6; in the meantime there are a few simple cases for which the equations already derived are sufficient. These relate to fields in which, from considerations of geometrical symmetry, we can deduce the distributions of charges over the conductors.

## 3.6 Simple systems

### 3.6.1 Spherical conductors in free space

In fig. 3.7($a$) let the spherical conductor have radius $r_0$ and charge $Q$. We wish to find the field strength at a point $P$, distant $r$ from the centre of the conductor $O$, where $r$ is not less than $r_0$. We assume that there are no other charged bodies in the vicinity so that, from conditions of symmetry, $Q$ is distributed uniformly over the conductor.

Construct an imaginary spherical surface of radius $r$, concentric with the conductor, and apply (3.21) to this surface. By symmetry $\boldsymbol{E}$ will be

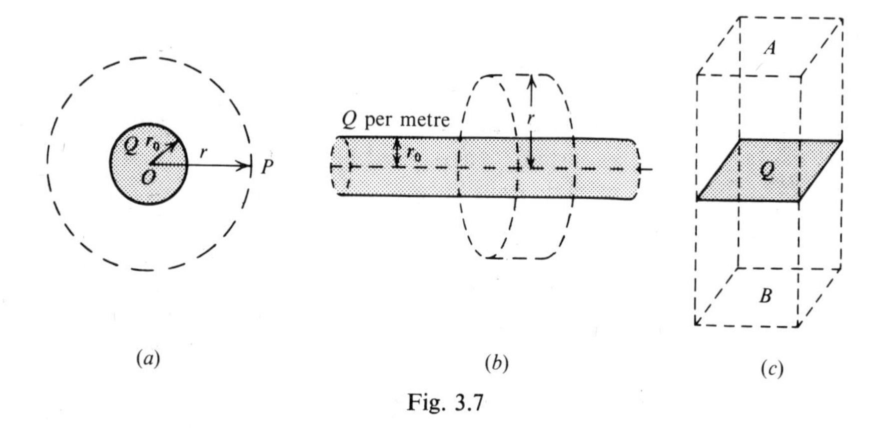


Fig. 3.7

constant over the surface and will be directed radially. Hence

$$\oint_S \boldsymbol{E}\cdot\boldsymbol{n}\,dS = 4\pi r^2 E = Q/\epsilon_0 \tag{3.29}$$

or

$$E = Q/4\pi r^2 \epsilon_0 \tag{3.30}$$

A surface such as the imaginary one that we have used in this problem is often known as a gaussian surface. It is to be noted that, so long as $r$ is not less than $r_0$ the field strength is the same as if the whole of the charge had been concentrated at the centre of the sphere.

### 3.6.2 Infinitely long cylindrical conductor

In fig. 3.7($b$) the cylindrical conductor of infinite length has radius $r_0$ and carries charge $Q$ per unit length, which is assumed to be uniformly distributed. By symmetry the field strength $\boldsymbol{E}$ at point $P$, distant $r$ from the axis, will be radial and we assume that $r$ is not less than $r_0$.

As a gaussian surface we take a cylinder of radius $r$, coaxial with the conductor, of unit axial length. There will be no flux of $\boldsymbol{E}$ passing through the plane sides of this surface so, applying (3.21)

$$\oint_S \boldsymbol{E}\cdot\boldsymbol{n}\,dS = 2\pi r E = Q/\epsilon_0 \tag{3.31}$$

and

$$E = Q/2\pi r \epsilon_0 \tag{3.32}$$

### 3.6.3 Infinite plane conductor

The conductor carries charge $Q$ per unit area and, by symmetry, the field everywhere must be in a direction normal to the plane. For a gaussian

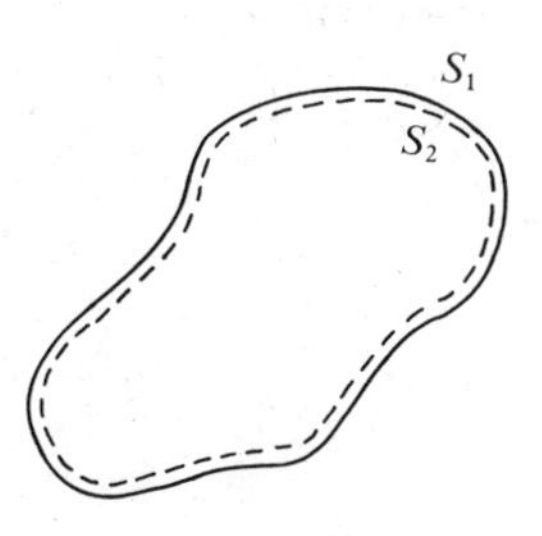


Fig. 3.8

surface, take a rectangular box enclosing unit area of the plane, as in fig. 3.7($c$). The whole of the flux of $\boldsymbol{E}$ will pass through the two faces $A$ and $B$ of the box and, by symmetry, the field strength at these two faces will be the same. Hence, by (3.21),

$$\oint \boldsymbol{E}\cdot\boldsymbol{n}\,\mathrm{d}S = 2E = Q/\epsilon_0 \tag{3.33}$$

and

$$E = Q/2\epsilon_0 \tag{3.34}$$

This result may be compared with (3.11).

### 3.6.4 Field inside a closed hollow conductor containing no charge

In fig. 3.8 let $S_1$ be a closed hollow conductor containing no charge. Then, if all external charges are at rest, $S_1$ will be an equipotential surface. If $S_1$ is a perfect conductor, this will be true even though currents are flowing in it.

If an electric field exists within $S_1$, it will be possible to construct an imaginary equipotential surface $S_2$, adjacent to $S_1$. Within the space between $S_1$ and $S_2$ the field **E** is either directed wholly outward from $S_2$ to $S_1$ or wholly inward from $S_1$ to $S_2$. By Gauss' theorem this cannot happen in either case, unless there is charge within $S_2$.

We therefore conclude that no field $\boldsymbol{E}$ can exist inside a closed hollow conductor that contains no charge.

### 3.6.5 Potential resulting from a point charge or a charged conducting sphere

In fig. 3.9($a$) we suppose the point charge $Q$ at $O$ to be isolated in free space at an infinite distance from all other charges, and we arbitrarily decide to measure potential with respect to a point at infinity. With this choice of zero, the potential at a point $P$, distant $r$ from $Q$, is the work done in bringing unit charge from infinity to $Q$. This work is independent of the

Fig. 3.9

path and we choose a path along the projection of $OP$. The force repelling the unit charge when it is distant $x$ from $O$ is $Q/4\pi\epsilon_0 x^2$, so the potential $V$ at $P$ is, by (3.22),

$$V = \int_r^\infty \frac{Q\,\mathrm{d}x}{4\pi\epsilon_0 x^2} = \frac{Q}{4\pi\epsilon_0 r} \tag{3.35}$$

From (3.30) this result holds also for a uniformly charged sphere of radius $r_0$, so long as $r$ is not less than $r_0$ (fig. 3.9(*b*)).

If we adopt a similar procedure to find the potential cased by an infinitely long charged cylinder or infinite charged plane, the result will be infinite in each case. This has no physical significance and means only that our choice of a point at infinity as the zero from which to measure potential is inappropriate in these cases. In neither case is there any difficulty in calculating the difference of potential between two points separated by a finite distance, as we shall see in §3.7.3 and §3.7.4.

### 3.6.6 The electric dipole

When a positive point charge $Q$ is separated from an equal negative point charge $-Q$ by a distance $l$, the combination of the two charges is known as an *electric dipole*. Dipoles play an important part in physical theory and we wish to derive expressions for the potential $V$ and field strength $\boldsymbol{E}$ which the dipole produces at a point $P$ distant $r$ from its centre. We are particularly interested in the case when $r$ is very large compared with $l$.

Since potential is a scalar quantity, its value at $P$ will be the sum of the values which each of the two charges, acting separately, would produce. Taking a point at infinity as our zero, we have, with the notation of fig. 3.10,

$$V = \frac{Q}{4\pi\epsilon_0}\left(\frac{1}{r_1} - \frac{1}{r_2}\right) = \frac{Q}{4\pi\epsilon_0}\left(\frac{r_2 - r_1}{r_1 r_2}\right) \tag{3.36}$$

If $r$ is very large compared with $l$, we may write

$$r_2 - r_1 = l\cos\theta, \quad r_1 r_2 = r^2 \tag{3.37}$$

and

$$V = Ql\cos\theta/4\pi\epsilon_0 r^2 \tag{3.38}$$

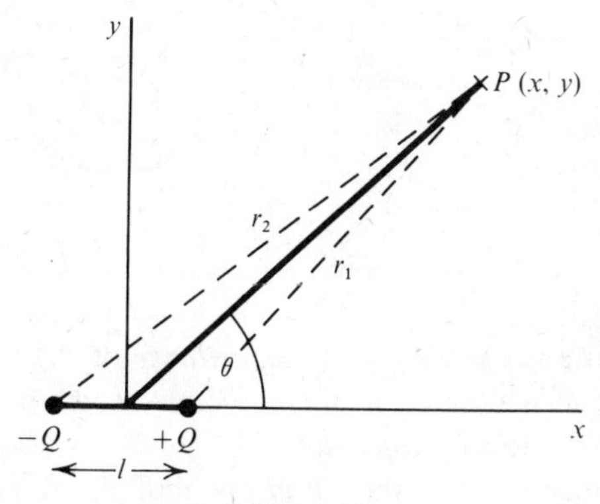


Fig. 3.10

The product of $Q$ and $l$ is known as the dipole moment $p$. If we wish to draw attention to the fact that the dipole has directional properties, we write

$$\boldsymbol{p} = Q\boldsymbol{l} \tag{3.39}$$

where the direction of $\boldsymbol{l}$ is from the negative to the positive charge. Then

$$V = \boldsymbol{p}\cdot\boldsymbol{r}_0/4\pi\epsilon_0 r^2 \tag{3.40}$$

where $\boldsymbol{r}_0$ is the unit vector along $r$.

We may find the components of the electric field strength $\boldsymbol{E}$ by differentiating (3.38). Thus, for the component along $r$, we have

$$E_r = -\frac{\partial V}{\partial r} = 2Ql\cos\theta/4\pi\epsilon_0 r^3 = p\cos\theta/2\pi\epsilon_0 r^3 \tag{3.41}$$

and for the component normal to $r$

$$E_\theta = -\frac{1}{r}\frac{\partial V}{\partial\theta} = Ql\sin\theta/4\pi\epsilon_0 r^3 = p\sin\theta/4\pi\epsilon_0 r^3 \tag{3.42}$$

## 3.7 Capacitance

### 3.7.1 Definition of capacitance

In fig. 3.11 let $A$ and $B$ be two conductors, of any shape or size, carrying charges $+Q$ and $-Q$ respectively, and let the dashed line represent an arbitrary path between the two conductors. At each point on this line there will be components of electric field strength caused by each of the two charges, and each of the components will be proportional to $Q$. Thus the resultant field strength $\boldsymbol{E}$ at every point on the line will be proportional to $Q$ and so also, by (3.22), will be the potential difference $V$ between $A$ and $B$. Thus we may write

$$Q = CV \tag{3.43}$$

Fig. 3.11

where $C$ is a constant known as the *capacitance* of the two conductors. A physical entity, such as the pair of conductors, which possesses capacitance, will be referred to as a *capacitor*.

The unit of capacitance is the *farad* (symbol F). A capacitor whose capacitance is one farad is one such that positive and negative charges of one coulomb respectively, on its conductors, cause a potential difference of one volt between those conductors.

### 3.7.2 Capacitance of concentric spherical conductors

In fig. 3.12($a$) $A$ represents a spherical conductor of radius $r_A$ and carrying charge $Q$, inside and concentric with a hollow spherical conductor $B$, of radius $r_B$ and carrylng charge $-Q$. The dashed line represents a spherical gaussian surface of radius $r$, concentric with the conductors. By symmetry, the field strength $\boldsymbol{E}$ over this surface will be constant and radial. Then, by (2.20)

$$\oint_S \boldsymbol{E}\cdot\boldsymbol{n}\mathrm{d}S = 4\pi r^2 E = Q/\epsilon_0 \tag{3.44}$$

and

$$E = Q/4\pi\epsilon_0 r^2 \tag{3.45}$$

We note that the negative charge on $B$ does not affect the value of $E$. By (2.22)

$$V_A - V_B = \frac{Q}{4\pi\epsilon_0}\int_A^B \frac{\mathrm{d}r}{r^2} = \frac{Q}{4\pi\epsilon_0}\left(\frac{1}{r_A}-\frac{1}{r_B}\right) \tag{3.46}$$

whence

$$C = Q/V = 4\pi\epsilon_0 r_A r_B/(r_B - r_A)\ \mathrm{F} \tag{3.47}$$

As a special case of (3.47) let $r_B$ increase without limit, while $r_A$ remains constant. We then have

$$C = 4\pi\epsilon_0 r_A\ \mathrm{F} \tag{3.48}$$

which may be taken to be the capacitance of an isolated spherical conductor suspended in free space. It is equal to the charge needed to raise the potential of the sphere by one volt, measured with respect to the potential of a point at infinity.

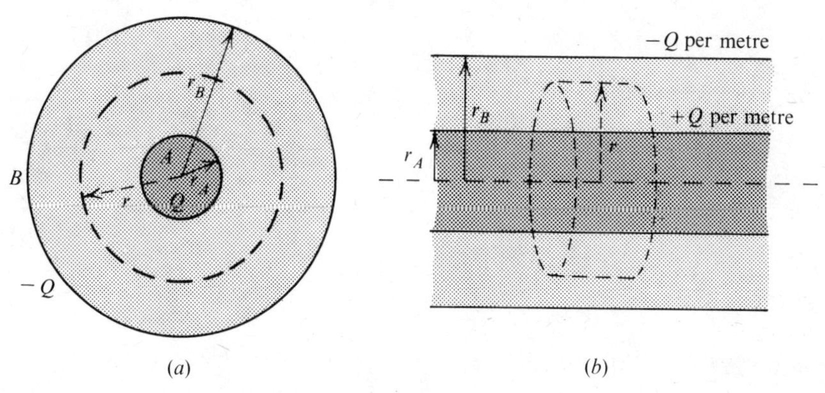


Fig. 3.12

### 3.7.3 Capacitance of coaxial cylinders

To avoid end effects we suppose the system to be infinitely long and consider a section of unit length which carries charges $Q$ and $-Q$ on the inner and outer cylinders respectively. The radii of these cylinders are $r_A$ and $r_B$, as shown in fig. 3.12(*b*).

The gaussian surface is a coaxial cylinder of radius $r$ and unit length. By symmetry, the electric field will be zero over the plane faces of this surface and will have a constant value $E$, in a radial direction, over the curved surface. Then

$$\oint_S \boldsymbol{E}\cdot\boldsymbol{n}\mathrm{d}S = 2\pi rE = Q/\epsilon_0$$

or

$$E = Q/2\pi\epsilon_0 r \tag{3.49}$$

Then

$$V_A - V_B = \frac{Q}{2\pi\epsilon_0}\int_A^B \frac{\mathrm{d}r}{r} = \frac{Q}{2\pi\epsilon_0}\ln\frac{r_B}{r_A} \tag{3.50}$$

and

$$C = Q/V = 2\pi\epsilon_0/\ln(r_B/r_A)\ \mathrm{F\,m^{-1}} \tag{3.51}$$

### 3.7.4 Infinite parallel plane conductors

Let the parallel plane conductors be separated by distance $d$. We suppose the conductors to be of infinite extent and consider unit area of the system carrying charges $Q$ and $-Q$. From the symmetry of the system, fig. 3.13(*a*), it is clear that the field between the plates is uniform, with constant value $E$ at right angles to the plates. It is generally assumed, as an obvious fact, that the whole of the electric flux leaving charge on $A$ ends on charge on $B$. We will later discuss this assumption; for the moment we accept it. Then, taking as gaussian surface the rectangular box shown in fig. 3.13(*a*), the

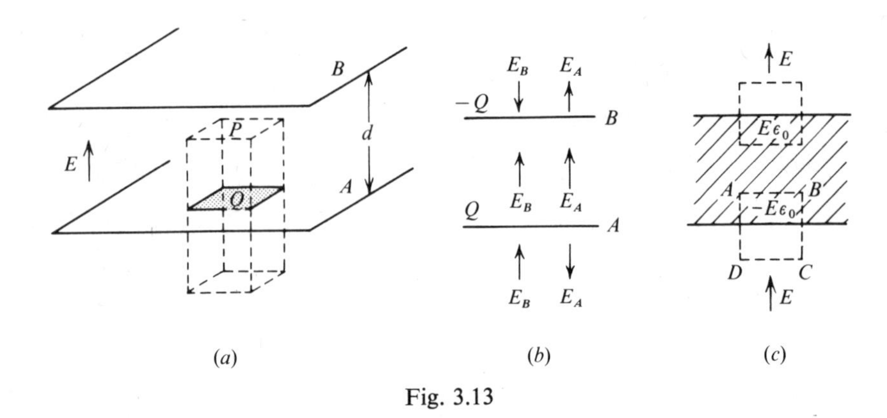


Fig. 3.13

whole of the flux from $Q$ passes through face $P$ and

$$\oint_S \boldsymbol{E}\cdot\boldsymbol{n}\,\mathrm{d}S = E = Q/\epsilon_0 \tag{3.52}$$

and $$V_A - V_B = Ed = Qd/\epsilon_0$$

whence $$C = \epsilon_0/d \text{ F m}^{-2} \tag{3.53}$$

The reader may feel disinclined to accept the assumption made above since, when we discussed the field caused by a single charged plane conductor (§3.6.3), we stated that half of the flux would emerge from each side of the plate, giving rise to field strength

$$E = Q/2\epsilon_0 \tag{3.54}$$

Accepting this result, we must say that, by the principle of superposition, the total field resulting from the charges on the two isolated plates is the sum of the fields which each, acting separately, would produce. The components acting in the various regions of the field are then as shown in fig. 3.13(*b*) and each of the components $E_A$ and $E_B$ is equal to $Q/2\epsilon_0$. Between the conductors these components add to give the result of (3.52) while, outside the conductors, they cancel to give zero field.

Once again the reader may feel that this result conflicts with a preconceived idea that a conducting sheet acts as a screen for an electrostatic field, so we examine the matter a little further. Let fig. 3.13(*c*) represent a portion of a single isolated uncharged conducting sheet whose thickness is finite and let a uniform electric field $E$ be incident normally on its lower surface. We take a gaussian surface in the form of a rectangular box, represented in section by $ABCD$ which intersects unit area of the surface of the conductor. By symmetry there is no flux of $\boldsymbol{E}$ across the vertical sides of the box and, because the interior of the conductor is an equipotential region, there is no flux crossing the side $AB$. Thus the total flux entering the box is that passing through the unit area $DC$ and this is equal to $E$. We therefore deduce (cf. (3.52)) that there must be a charge equal to $-E\epsilon_0$ on that part of the conductor within the box. The conductor was initially uncharged, so any charge near its lower surface must have left an equal and opposite charge near its upper surface and, by a reversal of the previous argument, this latter charge will give rise to a field $E$ in the region above the upper surface. What has happened is that the mobile charge within the conductor has redistributed itself to ensure that there shall be constancy of potential in this region, but this has not affected the field outside the conductor. It is to be emphasized that this result applies to an isolated conducting sheet; if the sheet is connected to earth, a large extra conductor has, in effect, been added to the system and we must expect this to cause changes in the electric fields. This matter will arise again in §§3.7.6 and 3.7.7.

Finally, for those who do not wish to follow the above arguments in detail, we offer the following alternative derivation of the capacitance of parallel plane conductors. For concentric spheres we have derived the relation (3.47)

$$C = 4\Sigma\pi_0 r_A r_B/(r_B - r_A) \tag{3.55}$$

Let both $r_A$ and $r_B$ get very large, with a mean value $r$, while $(r_B - r_A)$ remains constant and has the value $d$. Then (3.55) becomes

$$C = 4\pi r^2 \epsilon_0/d \tag{3.56}$$

In the limit, as $r$ tends to infinity, any portion of the system will be indistinguishable from parallel planes, and the capacitance per unit area will be

$$C = \epsilon_0/d \text{ F m}^{-2}$$

in agreement with (3.54).

### 3.7.5 Infinitely long parallel cylindrical conductors

The capacitance between very long parallel cylindrical conductors is of practical importance because conductors of this form are widely used in communication engineering; for example, in overhead telephone lines. A direct solution of this problem, using the methods so far discussed, is not possible, but we can obtain the capacitance per unit length by an artifice, as follows.

We begin by finding the potential at a point distant $r$ from an infinitely long linear charge, of magnitude $Q$ per unit length. This is a case where the choice of a point at infinity as our zero of potential is unsatisfactory, since it leads to an infinite potential near the conductor. As explained earlier,

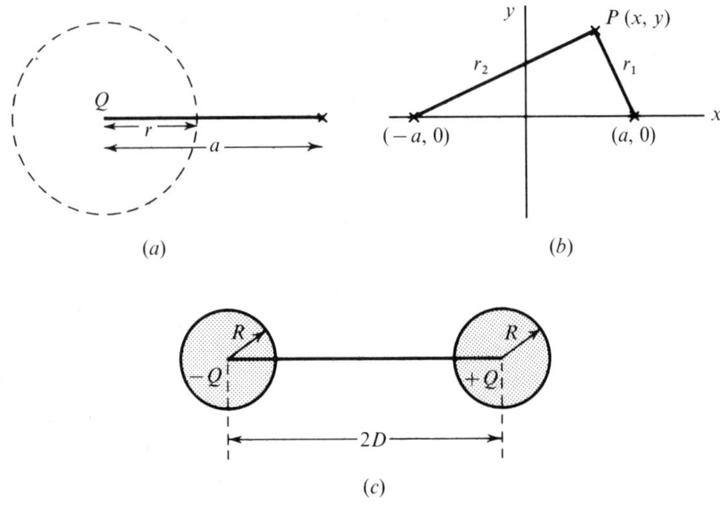


Fig. 3.14

this is of no physical significance and, for reasons which will shortly appear, we take our zero to be the potential at a point distant $a$ from the linear charge, as in fig. 3.14($a$). From a gaussian surface of radius $r$, it is apparent that the field strength at distant $r$ from the charge is

$$E = Q/2\pi\epsilon_0 r$$

and is directed radially away from the line. Thus, the potential at distance $r$, which is equal to the work that must be done in bringing unit charge from distance $a$ to distance $r$, is

$$V = -\frac{Q}{2\pi\epsilon_0}\int_a^r \frac{dr}{r} = \frac{Q}{2\pi\epsilon_0}\ln\frac{a}{r} \tag{3.57}$$

Next, consider two infinite line charges, with $+Q$ and $-Q$ per unit length respectively, parallel to the $z$-axis and cutting the $xy$-plane at $(a, 0)$ and $(-a, 0)$ respectively, as in fig. 3.14($b$). From the symmetry of the system the $yz$-plane is an equipotential and we take this as our zero of potential. From (3.57) the potential of any point $P$, $(x, y)$, is

$$V_P = \frac{Q}{2\pi\epsilon_0}\left(\ln\frac{a}{r_1} - \ln\frac{a}{r_2}\right) = \frac{Q}{2\pi\epsilon_0}\ln\frac{r_2}{r_1} \tag{3.58}$$

Thus, the surfaces for which $r_2/r_1 = k$, where $k$ is a constant, will be equipotentials. These surfaces cut the $xy$-plane in curves whose equations are given by

$$k^2 = [(a+x)^2+y^2]/[(a-x)^2+y^2]$$

or

$$x^2+y^2-[2ax(k^2+1)/(k^2-1)] = -a^2$$

Hence these curves are circles whose radii depend on the values of $k$. For any one value, the radius $R$ is given by

$$R = 2ak/(k^2-1) \tag{3.59}$$

and the centres of the two corresponding circles lie on the $x$-axis at the points $(D, 0)$ and $(-D, 0)$, where

$$D = [a(k^2+1)/(k^2-1)] \tag{3.60}$$

Now, in any electrostatic field, the surface of a conductor is an equipotential. Thus, if we place two infinite rods of radius $R$ to coincide with the equipotentials of this radius and if the charges $+Q$ and $-Q$ per unit length are assumed to reside on these rods, the field outside the rods will not be altered. We now have the information required to calculate the capacitance $C$ per unit length between the rods (fig. 3.14($c$)).

From (3.58), the total potential difference between the rods will be

$$2V_P = (Q \ln k)/\pi\epsilon_0 \tag{3.61}$$

and

$$C = Q/2V_P = \pi\epsilon_0/\ln k \tag{3.62}$$

To express $k$ in terms of the radii and axial separation of the rods, we have from (3.59) and (3.60)

$$D/R = (k^2+1)/2k$$

whence

$$k = [D \pm \surd(D^2-R^2)]/R \tag{3.63}$$

The positive sign is appropriate for the cylinder on the right of the figure, where $Q$ and $D$ are both positive; the negative sign leads to the same result for the left-hand cylinder, with both $Q$ and $D$ negative. In both cases

$$C = \pi\epsilon_0/\ln \{[D+ \surd(D^2-R^2)]/R\} \tag{3.64}$$

It is left to the reader to show that this equation can also be expressed in the form

$$C = \pi\epsilon_0/\cosh^{-1}(D/R) \tag{3.65}$$

### 3.7.6 Systems of several conductors

We have seen in the preceding section that the field strength in the region between two conductors may be affected by other charged conductors in the neighbourhood. We might therefore wonder whether the capacitances that we have been calculating for various pairs of conductors have any meaning unless the conductors under consideration are infinitely remote from all other charged bodies. We now examine a system which contains any number of conductors of arbitrary sizes and shapes.

Let the charges on the conductors be $Q_1, Q_2, \ldots, Q_n$. We suppose these charges to have been brought to the conductors from a point infinitely remote from the system and we arbitrarily decide to measure potential from this remote point. Let the potentials of the conductors then be $V_1, V_2, \ldots, V_n$. Because of the superposition theorem we can write

$$
\begin{aligned}
V_1 &= p_{11} Q_1 + p_{12} Q_2 + \ldots + p_{1n} Q_n \\
V_2 &= p_{21} Q_1 + p_{22} Q_2 + \ldots + p_{2n} Q_n \\
&\ldots\ldots\ldots\ldots\ldots\ldots\ldots\ldots \\
V_n &= p_{n1} Q_1 + p_{n2} Q_2 + \ldots + p_{nn} Q_n
\end{aligned}
\tag{3.66}
$$

where the $p$s are constants known as the *coefficients of potential.* They depend only on the geometry of the system. These linear equations can be solved to give the charges as functions of the potentials, in the form

$$
\begin{aligned}
Q_1 &= c_{11} V_1 + c_{12} V_2 + \ldots + c_{1n} V_n \\
Q_2 &= c_{21} V_1 + c_{22} V_2 + \ldots + c_{2n} V_n \\
&\ldots\ldots\ldots\ldots\ldots\ldots\ldots\ldots \\
Q_n &= c_{n1} V_1 + c_{n2} V_2 + \ldots + c_{nn} V_n
\end{aligned}
\tag{3.67}
$$

Since the $c$s depend only on the $p$s, the former are also constants depending only on the geometry of the system. The terms $c_{11}, c_{22}, \ldots, c_{nn}$ are known as *coefficients of capacitance,* while $c_{12}, c_{13}, \ldots, c_{jk}$ are termed *coefficients of induction.*

Let all the bodies other than 1 be initially uncharged and let us calculate the work done in putting charge $Q_1$ on 1. At any stage of this process, when the charge on 1 is $Q$, its potential is $p_{11} Q$ and the work done in adding a small additional charge $\mathrm{d}Q$ is $p_{11} Q\,\mathrm{d}Q$. Then the total work for charge $Q_1$ is

$$W_1 = \int_0^{Q_1} p_{11} Q\,\mathrm{d}Q = \tfrac{1}{2} p_{11} Q_1^2 \tag{3.68}$$

Keeping 1 charged let us find the extra work needed to put charge $Q_2$ on 2. When the charge on 2 is $Q$ its potential is $p_{21} Q_1 + p_{22} Q$ and the work to add further charge $\mathrm{d}Q$ is $(p_{21} Q_1 + p_{22} Q)\,\mathrm{d}Q$. Thus the total work to add $Q_2$ is

$$W_2 = \int_0^{Q_2} (p_{21} Q_1 + p_{22} Q)\,\mathrm{d}Q = p_{21} Q_1 Q_2 + \tfrac{1}{2} p_{22} Q_2^2 \tag{3.69}$$

and the total work to charge both 1 and 2 is

$$W = W_1 + W_2 = \tfrac{1}{2} p_{11} Q_1^2 + p_{21} Q_1 Q_2 + \tfrac{1}{2} p_{22} Q_2^2 \tag{3.70}$$

Clearly this total work cannot depend on whether 1 or 2 was charged first.

If we had begun with 2 we should have found

$$W = \tfrac{1}{2}p_{11}Q_1^2 + p_{12}Q_1Q_2 + \tfrac{1}{2}p_{22}Q_2^2 \tag{3.71}$$

We therefore conclude that $\quad p_{12} = p_{21}$

and in general,

$$p_{jk} = p_{kj} \tag{3.72}$$

From purely algebraic considerations it then follows that, in (3.67),

$$c_{jk} = c_{kj} \tag{3.73}$$

We now wish to re-write (3.67) in a rather different form, putting

$$C_{jj} = c_{j1} + c_{j2} + \ldots + c_{jn} \tag{3.74}$$

$$C_{jk} = -c_{kj} \tag{3.75}$$

and noting that $C_{jk}$ will be positive since, from algebraic considerations, $c_{jk}$ is negative. We can now write

$$\left.\begin{aligned} Q_1 &= C_{11}V_1 + C_{12}(V_1 - V_2) + \ldots + C_{1n}(V_1 - V_n) \\ Q_2 &= C_{21}(V_2 - V_1) + C_{22}V_2 + \ldots + C_{2n}(V_2 - V_n) \\ &\cdots\cdots\cdots\cdots\cdots\cdots\cdots\cdots \\ Q_n &= C_{n1}(V_n - V_1) + C_{n2}(V_n - V_2) + \ldots + C_{nn}V_n \end{aligned}\right\} \tag{3.76}$$

From (3.73) and (3.75) it follows that

$$C_{jk} = C_{kj} \tag{3.77}$$

Let us now concentrate our attention on conductors 1 and 2. We see that, of the total charge on 1, there is a component $C_{12}(V_1 - V_2)$ which is proportional to the potential difference between 1 and 2. Similarly, on 2, there is a component $C_{21}(V_2 - V_1)$ which is also proportional to this potential difference and which, by virtue of (3.77) is equal, and of opposite sign, to the component on 1. Moreover, from the preceding theory, it follows that $C_{12}$ and $C_{21}$ depend only on the sizes, shapes and positions of all the conductors in the system and are independent of the charges on these conductors. Thus it is legitimate to refer to $C_{12}$ as the capacitance between 1 and 2, knowing that this quantity is a constant geometrical property of the system.

In the above discussion we have taken a point at infinity as our zero of potential and have assumed the charges to have been brought to the conductors from this point. However, these are entirely arbitrary choices which do not in any way affect the argument. Let us consider the more practical system where conductor 1 is the earth, represented by a plane perfectly conducting sheet, and where there are several more charged

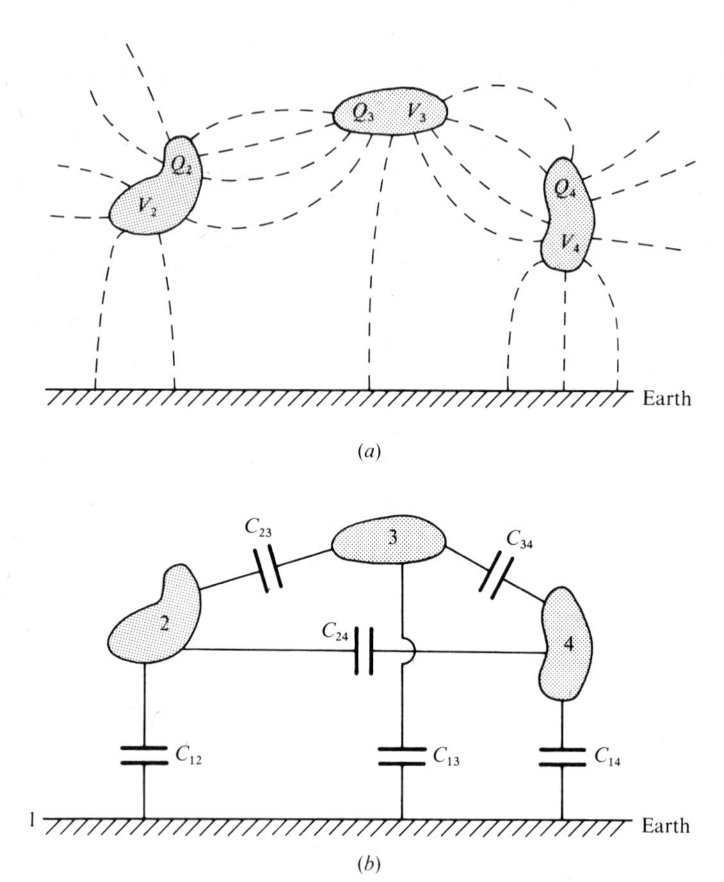


Fig. 3.15

conductors of which three are shown in fig. 3.15(*a*). Potentials may be specified with respect to the earth and the charges on the other conductors may be taken to have come from the earth.

The flux of $\boldsymbol{E}$ leaving any positive charge $Q$ or ending on any negative charge $-Q$ is $Q/\epsilon_0$. Thus, since we are taking $V_1$ to be zero, $C_{12}V_2/\epsilon_0$ is a measure of the flux passing from 2 to earth, $C_{23}(V_2-V_3)/\epsilon_0$ that passing between 2 and 3, and so on. For many purposes, particularly in network theory, it is convenient to represent the capacitances between the conductors by lumped capacitors, as in fig. 3.15(*b*).

To end, let us summarize what has been proved and what has not been proved in this section. For this purpose we consider the effect of conductor 4 and its charge on the interaction between 2 and 3.

We have not proved that $C_{23}$ is unaffected by the presence of 4, even though this conductor is uncharged. The presence of any conductor, whether charged or not, affects the form of the electrostatic field because it ensures that a certain region of space shall be at a constant potential whereas, without the conductor, it would probably not have been. Thus $C_{23}$ depends on the sizes, shapes and positions of all conductors in the field. What we have proved is that it does not depend on the charges which the conductors carry.

We have not proved that the flux of $\boldsymbol{E}$ passing between, say, 2 and 3 is unaffected by charge on 4. In general, a change in the charge on any conductor will affect the form of the whole electrostatic field, and hence the potentials of all the conductors. Since $C_{23}$ is a constant, any change in $V_2$ or $V_3$ which alters the difference $(V_2 - V_3)$ will change the flux passing between 2 and 3. It is sometimes said that the presence of a charge on 4 produces induced charges on 2 and 3, but we find this statement misleading. The total charges on 2 and 3, which are isolated conductors cannot be altered without actual contact with these conductors, though the distributions of the charges over the surfaces of the conductors can be changed. What the charge on 4 can do however is to alter the potentials of 2 and 3.

### 3.7.7 Shielding and earthing

We have already seen (§3.6.4) that there can be no electric field inside a closed hollow conductor which contains no charge. Suppose as in fig. 3.16 we have two charged conductors $A$ and $B$ within a closed surface $S$, as well as other charged conductors, such as $D$, outside. There will now be a field inside $S$ and, in general, flux of $\boldsymbol{E}$ will pass between $A$ and $B$, and between each of these and $S$. At any point within $S$ the field strength $\boldsymbol{E}$ will be the resultant of two components, $\boldsymbol{E}_1$ caused by the charges on $A$ and $B$, and $\boldsymbol{E}_2$ caused by all external charges. We have seen, however, that $\boldsymbol{E}_2$ is zero and, because it is zero, it cannot have affected the distribution of charge on either $A$ or $B$. We therefore conclude that the total field $\boldsymbol{E}$ is exactly the same as it would have been if there had been no external charges. We may say that $A$ and $B$ have been shielded from the effects of external charges.

Whenever the operation of electrical apparatus is likely to be affected by external electric fields, it is common practice to shield it by surrounding it with a closed conducting box. If dials within the box have to be observed, a portion of the box can be replaced by metal gauze of fine mesh, which acts as a nearly perfect screen.

It is often said that the shielding box must be earthed but, under the conditions stated, this is not necessary. If it is not earthed, external charges will affect the potential of the box and its contents, but internal fields will

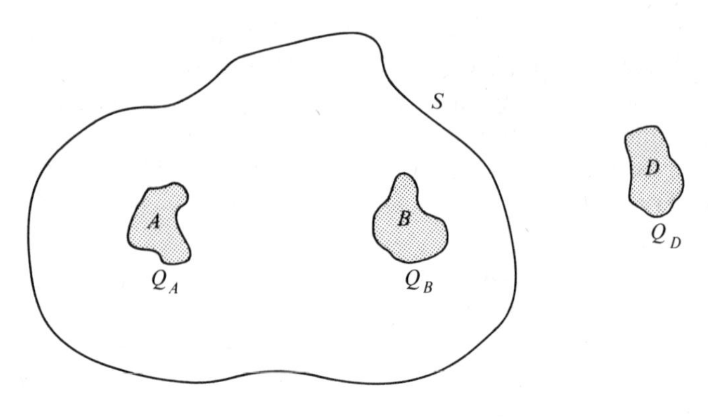


Fig. 3.16

not be altered. It often happens, however, that holes must be left in the shielding box, so that the operator can adjust equipment inside. It is then desirable that the box should always be at the same potential as the operator; earthing the box is a means to this end, though the potential of the operator may sometimes be appreciably different from earth, if he is standing on an insulating floor. It is by no means unknown for one end of a laboratory bench to be at a different potential from the other, as a result of minute leakages of current from the electric supply system. If the operator wishes to bring himself to the potential of the screen by touching it, he must avoid risk of shock by first connecting the screen to a conductor which is known to be at, or very near, earth potential (e.g. a water pipe or the earth wire of a supply system).

It is to be emphasized that the above discussion is concerned only with electrostatic effects. In the construction of electronic equipment it is standard practice to earth certain parts of the apparatus both to ensure the safety of the operator and to avoid interference from the a.c. supply system. These problems involve quite different considerations, which are outside the scope of this book.

### 3.7.8 Capacitors: guard rings

Capacitors are important components in electrical equipment and, in most applications, it is desirable that the capacitance between the two conductors should be definite and unaffected by their surroundings. From the preceding section, we see that this result can be achieved if one of the conductors almost completely encloses the other, leaving an opening only just large enough to allow contact to be made to the inner conductor.

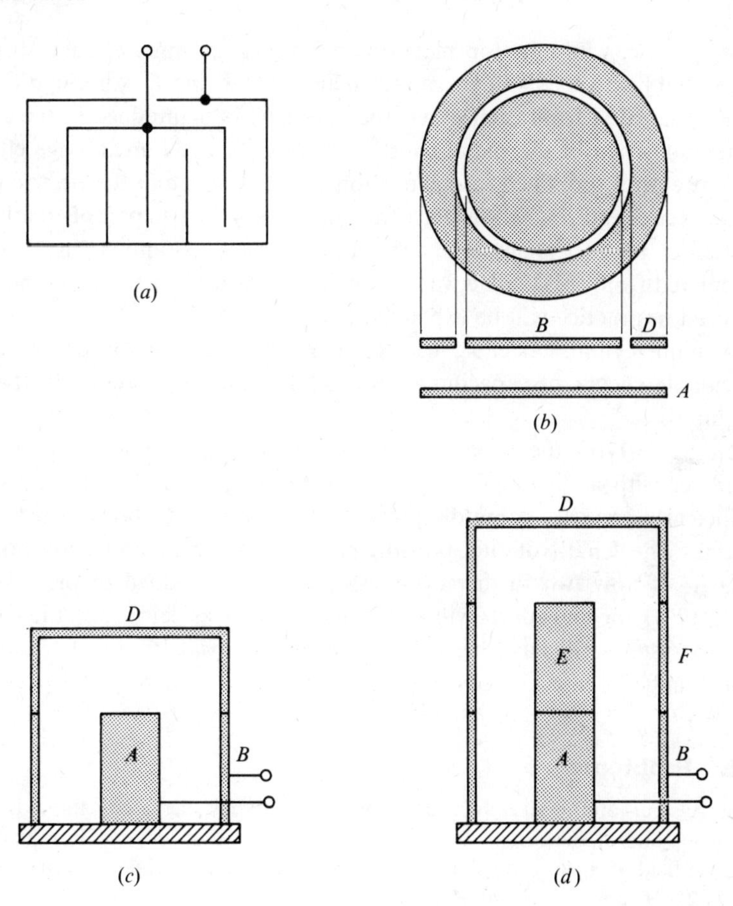


Fig. 3.17

A schematic arrangement for a parallel-plate capacitor embodying this principle is shown in fig. 3.17(*a*).

For certain purposes to be discussed later (§10.2.3), we need capacitors whose capacitances can be accurately calculated from their mechanical dimensions. Of the systems considered previously (§§3.7.2, 3.7.3 and 3.7.4), concentric spheres will not be considered further, since this arrangement is mechanically inconvenient. Our calculations for coaxial cylinders and parallel planes both refer to infinite systems; if finite portions of such systems are used, errors will be introduced as a result of edge effects. These effects are not very readily calculable and it is usual to eliminate them by the use of *guard rings*.

Taking first the parallel-plate capacitor, consider the arrangement shown in fig. 3.17(*b*), where the plates are circular. The bottom plate *A* is a single circular disc, while the top plate has two parts: an inner circular disc $B$, separated by a very small air gap from an annular ring $D$. Although $B$ and $D$ are insulated from each other, they are always maintained at the same potential. Then, if the radial width of $D$ is sufficiently great, edge effects will be entirely confined to this portion of the system and the electric field $\boldsymbol{E}$ between $B$ and $A$ will be exactly the same as if $B$ formed part of an infinite system of parallel plates. The capacitance between $B$ and $A$ can then be calculated from (3.53). The way in which an arrangement of this kind can be used in practice will be explained later (§7.3.2).

A finite cylindrical capacitor is an excellent device for providing a calculable *change* of capacitance, which for many purposes is all that is required.

In fig. 3.17(*c*) the capacitor consists of coaxial cylinders $A$ and $B$, together with a shielding cap $D$. The axial lengths of $A$ and $B$ must be sufficiently great for a middle portion of the system to be free from end effects. The length of this portion need not be large and need not be known. If now two further cylinders $E$ and $F$ are added to the system, fig. 3.17(*d*), the end effects will be exactly the same as before and the length of the system will have been increased by the axial length of $E$ and $F$. The change in capacitance can thus be calculated from (3.51).

## 3.8 Problems

1. A semi-infinite straight rod, of negligible radius, lies along the axis of $x$ with one end at the origin. It carries charge $Q$ per unit length. Show that the electric field at a point on the $y$-axis, distant $b$ from the origin, has magnitude $Q/2(\sqrt{2})\pi b\epsilon_0$ and is inclined at 45° to the $y$-axis.

2. Show that the solid angle included by a right circular cone of semi-angle $\alpha$ is $2\pi(1-\cos\alpha)$.

3. A charge $Q$ is situated at the point (3, 4) in the $xy$-plane and a charge $2Q$ is at the point (4, −3). Write down an expression for the potential, relative to a point at infinity, at any point $(x, y)$ in the plane. By differentiation of this expression, find the $x$-component of the field strength at the origin resulting from these two charges.

4. An infinitely long cylindrical rod of radius $r_0$ carries charge $Q$ per unit length. Show that the expression

$$E_r = -\partial V/\partial r$$

can be used to find the field strength at a point distant $r$ from the axis, where $r > r_0$.

5. A hollow conductor, made of thin metal, is closed except for a small hole in its surface. It is charged, but has no isolated charges inside it. Show that the electric field in the hole is $\sigma/2\epsilon_0$, where $\sigma$ is the surface density of charge in the region of the hole. (Hint: use the principle of superposition.)

6. Two concentric hollow spherical conductors in free space have radii $x$ and $y$ metres respectively ($x < y$). The inner sphere carries charge $+3Q$ coulombs and the outer one $-2Q$ coulombs.

(*a*) What is the potential difference between the conductors?

(*b*) What is the field strength at a point $z$ cm ($z > y$) from the centre?

7. A long straight cylindrical conductor of radius 1 mm lies along the axis of a long straight hollow conductor of square cross-section and of internal length of side 10 cm. Estimate the capacitance per metre length of this system.

8. Two long coaxial cylindrical conductors have a fixed potential difference $V$ applied between them. The external radius of the inner conductor is $R_1$ and the internal radius of the outer conductor is $R_2$. If $R_2$ is fixed, show that the field strength at the surface of the inner conductor will be a minimum when $R_2/R_1 = \mathrm{e}$ (the exponential coefficient).

# 4

# The magnetic field in free space

## 4.1 The basic postulate

### 4.1.1 Introduction

The study of magnetism may be said to have begun with the discovery of lodestone, an oxide of iron occurring in nature in the magnetized state, which was known to the Greeks. Later, the Chinese invented the magnetic compass and, in 1600, William Gilbert studied the properties of magnets in some detail.

These early experiments suggested that, in a bar magnet whose length was large compared with the dimensions of its cross-section, the sources of the magnetic properties lay in two *poles*, which were located near the ends of the bar. Gilbert noted that these two poles were different in kind, one being north-seeking and the other south-seeking, if the bar were freely suspended. He also noted that, when two magnets were in proximity, like poles repelled and unlike poles attracted each other. In 1785 Coulomb, experimenting with his torsion balance, enunciated a law for the force between magnetic poles which was precisely similar to the law which he had stated for the force between electric charges (§3.1.1). Based on this law, a theory of magnetostatics grew up, which had many parallels with electrostatics and, until a few decades ago, it was common for textbooks on electromagnetism to begin with an account of this theory.

In 1820 Oersted showed that magnetic effects could be produced by an electric current flowing in a circuit and the nature of magnetic fields generated in this way was explored by Ampère. Since a great deal was already known about the magnetic fields produced by permanent magnets, it was natural for Ampère to express his results by stating the distribution of magnetic poles to which a current-carrying circuit is equivalent (§4.1.9).

Experience has shown that there is a fundamental difference between the sciences of electrostatics and magnetostatics. Positive and negative electric charges occur separately and we know that, for example, the negatively charged electron is one of the fundamental particles, which is a constituent of all matter. On the other hand, north and south magnetic poles always occur in pairs; isolated poles of one kind have never been found. If, therefore, a fundamental magnetic particle exists on an atomic scale, we conclude that it must take the form of a dipole; a pair of equal and opposite poles separated by an extremely small distance. However, we shall show later (§4.1.7) that, on a macroscopic scale, the magnetic effect of an atomic dipole is precisely the same as that of a small loop of current of atomic dimensions. It is therefore immaterial whether we take the basic particle in a magnetized body to be a dipole or a current loop. Ampère himself surmised that the magnetic effects of magnetized bodies might result from minute circulating currents, sometimes now referred to as amperian currents.

We thus have the situation where all magnetic effects, whether resulting from magnetized bodies or from current-carrying circuits in free space, can be described either in terms of distributions of dipoles or in terms of distributions of currents. Nothing is to be gained by using both types of description and, on grounds of convenience and simplicity, we choose the latter. All magnetic effects will be assumed to result from the flow of current and we shall see later that the properties of magnetized bodies fit naturally into this scheme.

The above brief note is not intended to be a summary of the history of the development of the science of electromagnetism. It records a few of the milestones and explains how the modern viewpoint has arisen.

### 4.1.2 The basic experimental laws

In chapter 3 we saw that the properties of an electrostatic field could be specified by stating, for each point of the field, the magnitude and direction of the electric field strength $\boldsymbol{E}$. For the magnetic field we need a corresponding vector quantity, which is termed the *magnetic induction* or the *magnetic flux density*; we shall use the latter term. This vector is denoted by $\boldsymbol{B}$ and its properties, which must rest on experiment, will emerge as we proceed (§4.3.1). Our first task is to learn how $\boldsymbol{B}$ can be calculated for any circuit carrying a known current.

In fig. 4.1, let $\mathrm{d}l$ be a small element of the circuit $C$, which carries current $I$, and let it be required to calculate the value of $\boldsymbol{B}$ which the circuit produces at some point $P$. If $r$ is the distance of $P$ from $\mathrm{d}l$ and $\theta$ the angle between $\mathrm{d}l$ and the radius vector from $\mathrm{d}l$ to $P$ we assert, as an experimental law, that the magnitude $\mathrm{d}B$ which this current element contributes to the flux density at $P$ is given by

$$\mathrm{d}B = (\mu_0 I \mathrm{d}l \sin\theta)/4\pi r^2 \tag{4.1}$$

The direction of $\mathrm{d}B$ is along a line perpendicular to the plane containing

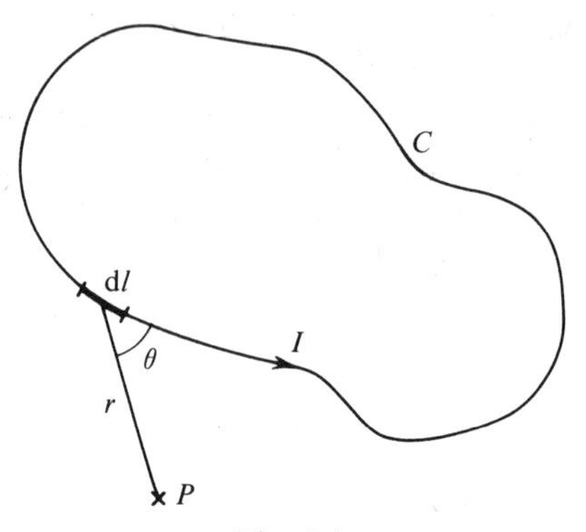


Fig. 4.1

both d$l$ and the radius vector $r$. Furthermore, if the radius vector were attached to a right-handed screw whose axis lay along d$l$, and if this screw were twisted in such a way as to cause it to move forward in the direction of $I$, the extremity of $r$ at $P$ would move in the direction of d$B$. Thus, in fig. 4.1, d$B$ is directed into the paper.

To find the total value of $\boldsymbol{B}$ at $P$, the components d$B$, calculated in accordance with the above rules for each element d$l$ of the whole circuit, must be added vectorially. It is important to notice that it is only the final value of $\boldsymbol{B}$ which can be verified by experiment, since there is no method by which we can produce an isolated element of current. Thus (4.1) is to be regarded as a true statement, only in the sense that it gives the correct result for a complete circuit. In (4.1), the $4\pi$ in the denominator of the right-hand side arises because SI units are rationalized, as explained in §3.1.2. The constant $\mu_0$ in the numerator is known as the *permeability of free space*, the *permeability of a vacuum*, or the *magnetic constant*; it has the exact value

$$\mu_0 = 4\pi \times 10^{-7}\ \mathrm{H\ m^{-1}}$$

This value is a consequence of the definition of the ampere, as will be shown later (§4.3.2). The unit for $\mu_0$ is discussed in §4.1.5.

The unit of magnetic flux density is the *tesla* (symbol T). An equivalent unit which is sometimes used is the *weber metre*$^{-2}$ (symbol Wb m$^{-2}$).

The above discussion has been concerned with flux density arising from the flow of current in a single circuit. When several currents, flowing in several separate circuits, are present in the field, experiment shows that the total magnetic flux density at any point is the vector sum of the flux densities which the separate currents would produce, if each were acting alone. We therefore adopt this *law of superposition* as a second basic postulate resting entirely on experimental evidence.

### 4.1.3 The vector form of the basic postulate *3*

The experimental law described in the previous section cannot be stated concisely in the vector notation that we have so far introduced. We therefore define a new quantity the *vector product* or *cross product* of two vectors $\boldsymbol{A}$ and $\boldsymbol{B}$, denoted by $\boldsymbol{A}\times\boldsymbol{B}$ or $\boldsymbol{A}\wedge\boldsymbol{B}$, and read '$\boldsymbol{A}$ cross $\boldsymbol{B}$'. As its name indicates, the vector product $C$ is itself a vector. It has magnitude

$$C = AB\sin\theta \tag{4.2}$$

where $\theta$ is the angle between $\boldsymbol{A}$ and $\boldsymbol{B}$. In direction it lies along the perpendicular to the plane containing $\boldsymbol{A}$ and $\boldsymbol{B}$ and is such that a right-handed screw with axis along this perpendicular would move in the direction of $\boldsymbol{C}$ when turning from $\boldsymbol{A}$ to $\boldsymbol{B}$. Thus, in the equation

$$\boldsymbol{C} = \boldsymbol{A}\times\boldsymbol{B} \tag{4.3}$$

the order in which $\boldsymbol{A}$ and $\boldsymbol{B}$ are written is important and

$$\boldsymbol{A}\times\boldsymbol{B} = -\boldsymbol{B}\times\boldsymbol{A} \tag{4.4}$$

With this new notation we can express (4.1) as

$$\mathrm{d}\boldsymbol{B} = \frac{\mu_0 I}{4\pi r^2}\,\mathrm{d}\boldsymbol{l}\times\boldsymbol{r}_0 \tag{4.5}$$

where $\boldsymbol{r}_0$ is the unit vector from the element of circuit to the point $P$, in fig. 4.1. This new statement is complete, in that it gives the direction of $\mathrm{d}\boldsymbol{B}$ as well as its magnitude. For the whole circuit we write

$$\boldsymbol{B} = \frac{\mu_0 I}{4\pi}\oint_l \frac{\mathrm{d}\boldsymbol{l}\times\boldsymbol{r}_0}{r^2} \tag{4.6}$$

where the integral sign instructs us to add vectorially the components of $\boldsymbol{B}$ arising from the various elements $\mathrm{d}\boldsymbol{l}$ and the circle on the integral sign denotes that the summation must be carried out round the complete circuit. Equation (4.6) does not tell us anything that was not stated in §4.1.2; it is merely a shorthand method of recording the experimental facts.

We shall adopt (4.6) as our basic postulate and as our definition of $\boldsymbol{B}$, in the study of the magnetic effects of currents. It is often known as the law of *Biot and Savart*, although these experimenters verified it only for current flowing in a long straight wire. It is also sometimes known as *Ampère's law*, since it is equivalent to, and can be derived from, the form in which Ampère first stated his results. The genius of Ampère lay in his elucidation of the relation between electric currents and magnetic fields; not in the accurate experimental verification of the law which he proposed. His experiments were necessarily crude by modern standards and the validity of his law rests on the vast number of accurate verifications of the consequences of the laws which have since been made. In general, these measurements can be related more directly to (4.6) than to the form of the law stated by Ampère himself.

### 4.1.4 Some properties of $\boldsymbol{B}$

It is a direct mathematical consequence of the definition of $\boldsymbol{B}$ contained in (4.6) that $\boldsymbol{B}$ is a flux vector. The proof of this statement will be given later (§4.1.9). Accepting its truth for the time being, we write

$$\phi = \int_S \boldsymbol{B}\cdot\boldsymbol{n}\,\mathrm{d}S \tag{4.7}$$

where $\phi$ is termed the *magnetic flux* or the *flux of* $\boldsymbol{B}$ through the surface $S$. The unit of magnetic flux is the *weber* (symbol Wb). As a result of the special property of a flux vector (§2.4.2), it follows that $\phi$ in (4.7) depends only on the perimeter of the surface $S$; it is the same for all surfaces having the same perimeter. It also follows that $\boldsymbol{B}$ can, in principle, be represented quantitatively by continuous lines, though we shall make little use of this concept. Examination of (4.6) shows that lines of $\boldsymbol{B}$ form closed loops encircling the current from which $\boldsymbol{B}$ results. Since we are adopting the standpoint that $\boldsymbol{B}$ always results from the flow of current, even in permanent magnets or magnetized bodies, lines of $\boldsymbol{B}$ will always form closed loops and it then follows that, for any closed surface $S$,

$$\oint_S \boldsymbol{B}\cdot\boldsymbol{n}\,\mathrm{d}S = 0 \tag{4.8}$$

If we had adopted the alternative view that, in magnetized bodies and permanent magnets, flux is to be ascribed to the existance of atomic magnetic dipoles, we should have said that lines of $\boldsymbol{B}$ begin on the north poles of these dipoles and end on their south poles. As we shall show (§4.1.9), the two descriptions are equivalent. Moreover, on a macroscopic scale, any closed surface will contain equal numbers of north and south poles, so (4.8) would still be true.

Although we shall ascribe no quantitative significance to magnetic poles, it is convenient to use the term qualitatively to indicate regions where flux is concentrated. Thus we may speak of the poles of a bar magnet or of the field magnet of a generator.

### 4.1.5 The magnetic field strength $\boldsymbol{H}$

If we were concerned only with magnetic effects in free space, resulting from the flow of current, the single vector $\boldsymbol{B}$ would be sufficient for the description of the magnetic field. However, when we come to the consideration of magnetized bodies, we shall find that a second vector is needed and it is convenient to introduce this vector forthwith. We therefore define the *magnetic field strength* $\boldsymbol{H}$, in free space, by the equation

$$\boldsymbol{B} = \mu_0 \boldsymbol{H} \tag{4.9}$$

or, from (4.6),

$$\boldsymbol{H} = \frac{I}{4\pi} \oint_l \frac{\mathrm{d}\boldsymbol{l} \times \boldsymbol{r}_0}{r^2} \tag{4.10}$$

$\boldsymbol{H}$ is sometimes known as the *magnetizing force*. It is measured in *amperes metre*$^{-1}$ (A m$^{-1}$).

Superficially there is a direct analogy between (4.9) and the relation

$$\boldsymbol{D} = \epsilon_0 \boldsymbol{E} \tag{4.11}$$

which was introduced in our discussion of electrostatic fields but, in fact, the situation is different in the two cases. In electrostatics $\boldsymbol{E}$ is the quantity which is physically measurable (e.g. by the force exerted on a known charge) and which directly specifies the properties of the field. $\boldsymbol{D}$ was introduced as a matter of mathematical convenience and was found to be related to the charges which produce the field, since unit flux of $\boldsymbol{D}$ begins on each unit positive charge and ends on each unit negative charge.

In the magnetic case, all physically observable effects depend directly on $\boldsymbol{B}$, which therefore describes the performance of the field. $\boldsymbol{H}$ is introduced as a matter of mathematical convenience and, by (4.10), is related to the current which produces the field. Thus the real analogy is between $\boldsymbol{E}$ and $\boldsymbol{B}$ on the one hand, and between $\boldsymbol{D}$ and $\boldsymbol{H}$ on the other. It follows from (4.10) that $\boldsymbol{H}$ has dimensions (current)(length)$^{-1}$ and we know that $\boldsymbol{B}$ has dimensions (flux)(length)$^{-2}$. Hence, from (4.9), $\mu_0$ has dimensions (flux)(current)$^{-1}$(length)$^{-1}$. We shall later (§4.5.1) introduce a unit of inductance, the *henry* (symbol H) and the reader may verify that it has dimensions (flux)(current)$^{-1}$. We can therefore take, as our SI unit for $\mu_0$, the *henry metre*$^{-1}$ (symbol H m$^{-1}$) and this is the unit commonly used.

### 4.1.6 The magnetic scalar potential *U*

In our discussion of the electrostatic field an important quantity was the potential $V$. We defined the difference of $V$ between two points in terms of the work done in carrying unit positive charge from one point to the other and showed that this definition led to the equations (2.34) and (2.35),

$$\boldsymbol{E} = -\frac{\partial V}{\partial n}\boldsymbol{n} = -\operatorname{grad} V \tag{4.12}$$

$$E_x = -\frac{\partial V}{\partial x}\boldsymbol{i}, \quad E_y = -\frac{\partial V}{\partial y}\boldsymbol{j}, \quad E_z = -\frac{\partial V}{\partial z}\boldsymbol{k}, \tag{4.13}$$

We now find it convenient to introduce a corresponding quantity for the magnetic field. We shall term it the *magnetic potential* (symbol $U$) and shall define it by the relation

$$\boldsymbol{H} = -\frac{\partial U}{\partial n}\boldsymbol{n} = -\text{grad } U \tag{4.14}$$

leading to the equations

$$H_x = -\frac{\partial U}{\partial x}\boldsymbol{i}, \quad H_y = -\frac{\partial U}{\partial y}\boldsymbol{j}, \quad H_z = -\frac{\partial U}{\partial z}\boldsymbol{k} \tag{4.15}$$

Strictly speaking $U$ should be termed the *magnetic scalar potential*, since a quite different quantity, the magnetic vector potential, will be introduced later. However, when there is no possibility of confusion, we shall refer to $U$ as the magnetic potential. It follows from (4.14) that the unit of magnetic potential is the *ampere*.

It is important to note that our definition of $U$ is a purely mathematical one in terms of $\boldsymbol{H}$ and that we have not made any assumptions about the properties of $U$ itself. In particular, we have *not* stated that the difference in $U$ between two points is equal to the work done in carrying unit magnetic pole from one point to the other. Such a statement would, in any case, be meaningless since isolated magnetic poles do not exist. Nor have we assumed that the difference in magnetic potential between two points is, under all circumstances, independent of the path followed from one point to the other.

$U$ has been introduced because it will prove to be useful; its properties must be deduced from our fundamental postulate expressed by (4.6).

### 4.1.7 Equivalence of a small plane current loop and a magnetic dipole

We have already stated (§4.1.1) that magnetic effects caused by magnetized bodies can be described equally well in terms of a distribution of atomic circulating currents or of a distribution of atomic magnetic dipoles. We now wish to show that a small plane current loop produces, at distant points, the same magnetic field strength $\boldsymbol{H}$ as one would expect from a magnetic dipole. We shall give two proofs of this equivalence, since both are instructive. In both proofs vector notation will be used since this greatly simplifies the arguments.

In fig. 4.2 current $I$ flows in a small plane circuit of which $\mathrm{d}\boldsymbol{l}$ is an element situated at $Q$. We wish to find the magnetic field strength at an arbitrary point $P$. The origin of coordinates is taken at a point $O$ within the circuit and the axes are orientated so that the circuit lies in the $xy$-plane

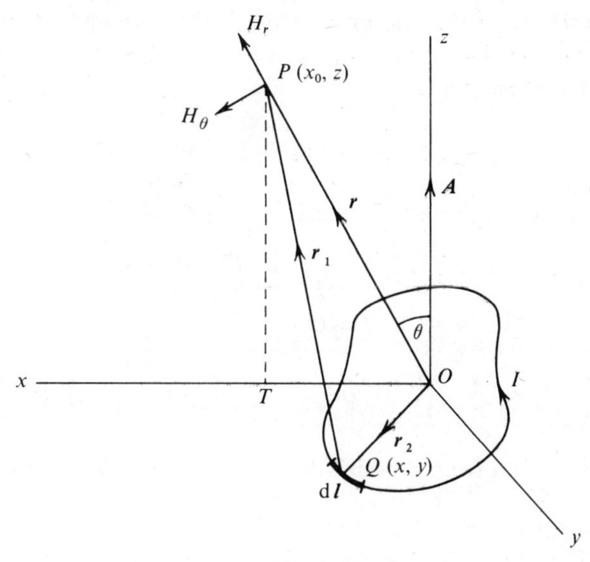


Fig. 4.2

and the $xz$-plane contains the line $OP$, which makes angle $\theta$ with the $z$-axis. The length $r$ of the vector $\boldsymbol{r}$ from $O$ to $P$ is large compared with the linear dimensions of the circuit. The coordinates of $Q$ are $(x, y)$ and those of $P$ are $(x_0, z)$. $\boldsymbol{r}_1$ is the vector from $Q$ to $P$ and $\boldsymbol{r}_2$ that from $O$ to $Q$. Then

$$r_1^2 = z^2+(x_0-x)^2+y^2$$

and

$$r^2 = z^2+x_0^2$$

Neglecting squares of small quantities,

$$r_1^2 = r^2-2xx_0 = r^2-2rx\sin\theta \tag{4.16}$$

The magnetic field strength at $P$ is to be calculated with the aid of (4.10) and, for the present purpose, it is convenient to write this equation in the form

$$\boldsymbol{H} = \frac{I}{4\pi}\oint_l \frac{\mathrm{d}\boldsymbol{l}\times\boldsymbol{r}_1}{r_1^3} \tag{4.17}$$

where, in accordance with fig. 4.2, the distance from d$\boldsymbol{l}$ to $P$ is denoted by $\boldsymbol{r}_1$ and the unit vector along $\boldsymbol{r}_1$ has been replaced by $\boldsymbol{r}_1/r_1$.

Since $r_1$ varies from point to point of the circuit, we express this length in terms of $r$, using (4.16). Neglecting squares and higher orders of small quantities,

$$\frac{1}{r_1^3} = \frac{1}{r^3}\left(1-\frac{2x}{r}\sin\theta\right)^{-\frac{3}{2}}$$

$$= \frac{1}{r^3}\left(1+\frac{3x}{r}\sin\theta\right) \tag{4.18}$$

With the directions of the vectors indicated in fig. (4.2), we have

$$\boldsymbol{r}_1 = \boldsymbol{r} - \boldsymbol{r}_2 \tag{4.19}$$

and, from (4.17) and (4.18),

$$\boldsymbol{H} = \frac{I}{4\pi r^3}\oint_l \left(1+\frac{3x}{r}\sin\theta\right)(\mathrm{d}\boldsymbol{l}\times\boldsymbol{r} - \mathrm{d}\boldsymbol{l}\times\boldsymbol{r}_2) \tag{4.20}$$

Since both $x$ and $r_2$ are small, we neglect the term containing their product and write

$$\boldsymbol{H} = \frac{I}{4\pi r^3}\oint_l \left[\mathrm{d}\boldsymbol{l}\times\boldsymbol{r} - \mathrm{d}\boldsymbol{l}\times\boldsymbol{r}_2 + \left(\frac{3x}{r}\sin\theta\right)\mathrm{d}\boldsymbol{l}\times\boldsymbol{r}\right] \tag{4.21}$$

Since $\boldsymbol{r}$ is a constant for the circuit, we may write (4.21) as

$$\boldsymbol{H} = \frac{I}{4\pi r^3}\left[\left(\oint_l \mathrm{d}\boldsymbol{l}\right)\times\boldsymbol{r} - \oint_l (\mathrm{d}\boldsymbol{l}\times\boldsymbol{r}_2) + \frac{3\sin\theta}{r}\left(\oint_l x\,\mathrm{d}\boldsymbol{l}\right)\times\boldsymbol{r}\right] \tag{4.22}$$

Taking the terms on the right-hand side one at a time, we note that

$$\oint \mathrm{d}\boldsymbol{l} = 0 \tag{4.23}$$

since the circuit is a closed curve. $\boldsymbol{r}_2\times\mathrm{d}\boldsymbol{l}$ is a vector parallel to the $z$-axis, of magnitude equal to the area of a parallelogram whose sides are $\boldsymbol{r}_2$ and $\mathrm{d}\boldsymbol{l}$. However, this area is twice the area of the triangle formed by $\mathrm{d}\boldsymbol{l}$ and the lines joining the ends of $\mathrm{d}\boldsymbol{l}$ to $O$. Thus we may write

$$-\oint \mathrm{d}\boldsymbol{l}\times\boldsymbol{r}_2 = \oint \boldsymbol{r}_2\times\mathrm{d}\boldsymbol{l} = 2\boldsymbol{A} \tag{4.24}$$

where $\boldsymbol{A}$ is a vector along the positive direction of the $z$-axis, of magnitude equal to the area of the circuit. Finally, we write

$$\mathrm{d}\boldsymbol{l} = \boldsymbol{i}\,\mathrm{d}x + \boldsymbol{j}\,\mathrm{d}y$$

where $\boldsymbol{i}$ and $\boldsymbol{j}$ are the unit vectors along the $x$- and $y$-axes respectively. Then

$$\oint_l x\,\mathrm{d}\boldsymbol{l} = \boldsymbol{i}\oint_l x\,\mathrm{d}x + \boldsymbol{j}\oint_l x\,\mathrm{d}y$$

The first integral on the right-hand side vanishes for the complete circuit, while

$$\boldsymbol{j}\oint_l x\,\mathrm{d}y = \boldsymbol{j}A \tag{4.25}$$

Collecting the terms and noting that

$$\boldsymbol{A} = \boldsymbol{k}A$$

where $\boldsymbol{k}$ is the unit vector along the $z$-axis, (4.22) becomes

$$\boldsymbol{H} = \frac{IA}{4\pi r^3}\left(2\boldsymbol{k} + \frac{3\sin\theta}{r}\boldsymbol{j}\times\boldsymbol{r}\right) \tag{4.26}$$

Taking components $H_r$ along $\boldsymbol{r}$ and $H_\theta$ at right angles to $\boldsymbol{r}$ in the direction of increasing $\theta$,

$$H_r = \frac{IA}{2\pi r^3}\cos\theta \tag{4.27}$$

$$H_\theta = \frac{IA}{4\pi r^3}\sin\theta \tag{4.28}$$

There is no component perpendicular to the $xz$-plane. The corresponding components of magnetic flux density are found by multiplying (4.27) and (4.28) by $\mu_0$ and the analogy between these equations and (3.41) and (3.42), which we derived for an electric dipole, are obvious. We therefore define a *magnetic dipole moment* $\boldsymbol{m}$ for the current loop by the equation

$$\boldsymbol{m} = \mu_0 I\boldsymbol{A} \tag{4.29}$$

Thus $\boldsymbol{m}$ has magnitude $\mu_0 IA$ and is directed along a line perpendicular to the plane of the loop in the sense that a right-handed screw would move if twisted in the direction of current flow. Then (4.27) and (4.28) can be written

$$H_r = m\cos\theta/2\pi\mu_0 r^3 \tag{4.30}$$

$$H_\theta = m\sin\theta/4\pi\mu_0 r^3 \tag{4.31}$$

in exact analogy with (3.41) and (3.42). Since, by our definition of magnetic potential $U$,

$$H_r = -\frac{\partial U}{\partial r} \tag{4.32}$$

we can integrate (4.30) to give

$$U = -\frac{m\cos\theta}{2\pi\mu_0}\int\frac{\mathrm{d}r}{r^3} = \frac{m\cos\theta}{4\pi\mu_0 r^2} + \text{const.} \tag{4.33}$$

The value of the constant of integration depends on our choice of zero from which to measure potential. If we choose a point at infinity, where the magnetic field strength must be zero, the constant vanishes and we have

$$U = \frac{m\cos\theta}{4\pi\mu_0 r^2} = \frac{\boldsymbol{m}\cdot\boldsymbol{r}_0}{4\pi\mu_0 r^2} = \frac{IA\cos\theta}{4\pi r^2} \tag{4.34}$$

A useful alternative form of this equation arises from the fact that $A\cos\theta/r^2$ is equal to the solid angle $\Omega$ which the current loop subtends at the point $P$. Thus

$$U = I\Omega/4\pi \tag{4.35}$$

We have now proved that, in every respect, the magnetic field at a distant point which results from the flow of current in a small plane loop is identical with the field that would have been caused by a magnetic dipole, of strength and orientation given by (4.29), on the assumption that

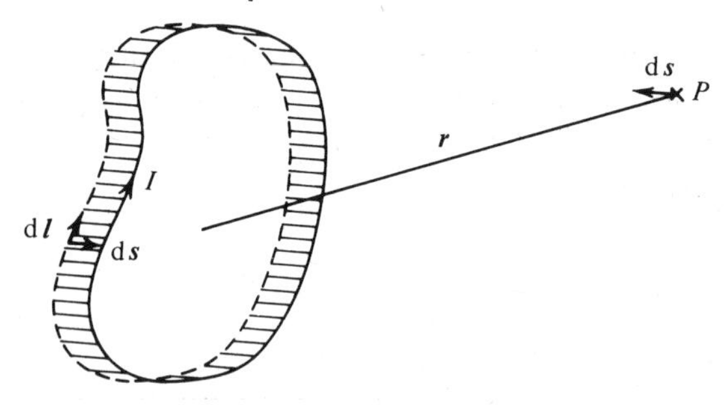


Fig. 4.3

the individual poles of the dipole obey laws similar to those which govern the properties of electrostatic charges. Thus, when we come to consider magnetized bodies, it will be immaterial whether we assume the magnetic properties of atoms or molecules to arise from circulating currents or permanent dipoles.

### 4.1.8 Alternative derivation of (4.35)

The relation between $U$, $I$ and $\Omega$ embodied in (4.35), which could hardly have been foreseen at the outset of our investigation, can now be derived by a simpler method.

In fig. 4.3 let current $I$ be flowing in a small plane loop and let it be required to calculate the magnetic potential $U$ at a point $P$, distant $\boldsymbol{r}$ from the loop, where $r$ is large in comparison with the dimensions of the loop. Let $\boldsymbol{H}$ be the magnetic field strength at $P$. If we move a small distance $\mathrm{d}\boldsymbol{s}$ from $P$, in any direction, the potential will change by $\mathrm{d}U$, where

$$\mathrm{d}U = -\boldsymbol{H}\cdot\mathrm{d}\boldsymbol{s} \tag{4.36}$$

$$= -I\mathrm{d}\boldsymbol{s}\cdot\oint_l \frac{\mathrm{d}\boldsymbol{l}\times\boldsymbol{r}}{4\pi r^3} = -\frac{I}{4\pi r^3}\oint_l \mathrm{d}\boldsymbol{s}\cdot(\mathrm{d}\boldsymbol{l}\times\boldsymbol{r}) \tag{4.37}$$

where $\mathrm{d}\boldsymbol{s}$ has been taken inside the integral, since it is constant at all points of the loop. Clearly the same change of potential would have occurred if there had been no displacement from $P$ but the current loop had been shifted through a distance $-\mathrm{d}\boldsymbol{s}$, as indicated in fig. 4.3. Moreover

$$\mathrm{d}\boldsymbol{s}\cdot(\mathrm{d}\boldsymbol{l}\times\boldsymbol{r}) = \boldsymbol{r}\cdot(\mathrm{d}\boldsymbol{s}\times\mathrm{d}\boldsymbol{l}) \tag{4.38}$$

since both sides of this equation represent the volume of a parallel-sided solid whose edges are $\mathrm{d}\boldsymbol{s}$, $\mathrm{d}\boldsymbol{l}$ and $\boldsymbol{r}$ respectively. Therefore, (4.37) becomes

$$\mathrm{d}U = -\frac{I}{4\pi r^3}\oint_l \boldsymbol{r}\cdot(\mathrm{d}\boldsymbol{s}\times\mathrm{d}\boldsymbol{l}) \tag{4.39}$$

Now $\mathrm{d}\boldsymbol{s} \times \mathrm{d}\boldsymbol{l}$ is the area swept out by the circuit element $\mathrm{d}\boldsymbol{l}$ during the displacement $-\mathrm{d}\boldsymbol{s}$ and $-\boldsymbol{r}\cdot(\mathrm{d}\boldsymbol{s} \times \mathrm{d}\boldsymbol{l})/r^3$ is the solid angle subtended by this area at $P$. Hence, integrating for the whole circuit,

$$\mathrm{d}U = I\mathrm{d}\Omega/4\pi \tag{4.40}$$

where $\mathrm{d}\Omega$ is the total change, caused by the displacement $-\mathrm{d}\boldsymbol{s}$, in the solid angle subtended by the circuit at $P$. If we take our zero of potential to be that at a point at infinity, the angle subtended by the circuit at this point will also be zero. Then, as $P$ is approached by any path, both $U$ and the solid angle will increase until, at $P$, we may write

$$U = I\Omega/4\pi \tag{4.41}$$

where $\Omega$ is the solid angle subtended at $P$ by the circuit.

### 4.1.9 Extension to a circuit of any size and shape

In the two preceding sections, our discussion has been limited to the magnetic field produced by a small plane coil at a point whose distance from the coil is large compared with the dimensions of the latter. To extend the results to a circuit of any size and shape, we use the artifice illustrated in fig. 4.4. Points on the circuit are joined by a number of conductors, which are themselves connected at their intersections, to form a network of small meshes. Current $I$ is assumed to flow round each of these small meshes and it is clear that, in all the conductors that have been added, the total current will be zero, leaving current $I$ flowing round the original circuit. Thus, by the principle of superposition, the magnetic potential at any point produced by the original current is equal to the sum of the potentials that would be produced by the same current flowing round each of the meshes in turn.

We can make the meshes as small as we please so that, whatever the size and shape of the original circuit and whether or not it lies in a plane, each mesh can be made effectively plane and of dimensions small compared with its distance from the point at which the magnetic potential is to be calculated. Thus, if $\delta U_1$, $\delta U_2$... are the components of the total potential and $\delta\Omega_1$, $\delta\Omega_2$... the solid angles subtended by the separate meshes,

$$U = \Sigma\delta U_n = I\Sigma\delta\Omega_n/4\pi = I\Omega/4\pi \tag{4.42}$$

where $\Omega$ is the total solid angle subtended at the point.

We have already seen, (4.29), that the magnetic field strength (and therefore the magnetic potential) at any point, caused by the flow of current $I$ in a small plane circuit of area $\delta A$, is equal to that which would

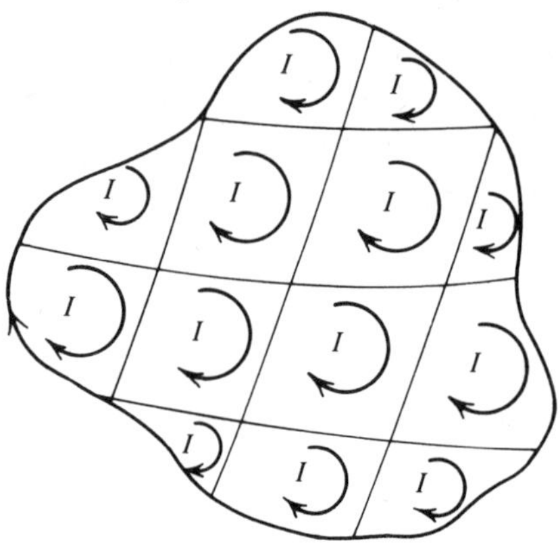


Fig. 4.4

have been produced by a magnetic dipole of moment $\boldsymbol{m}$, where

$$\boldsymbol{m} = \mu_0 I\delta\mathbf{A} \tag{4.43}$$

the dipole being situated at the circuit, with its axis at right angles to the plane of the circuit. In view of (4.42), we can now state that, so far as magnetic effects are concerned, a current $I$ flowing in a circuit of any size and shape is equivalent to a magnet in the form of a sheet of uniform thickness, whose boundary coincides with the circuit, which is magnetized at right angles to its surface at all points, and which has dipole moment $\mu_0 I$ per unit area. A magnet of this form, which it would be impossible to construct practically, is known as a *magnetic shell*. It will be appreciated that the equivalence of the current $I$ and the magnetic shell holds only for points outside the shell. However, this causes no difficulty because only the boundary of the shell is specified and an infinite number of shells can be imagined, all having the same boundary. Thus it is always possible to choose a shell which does not pass through the point at which we wish to calculate the magnetic field strength.

The fact that the magnetic field caused by a current $I$ is, at all points, the same as that which would be produced by the array of dipoles which we have termed a magnetic shell, leads to a result of great importance. In our discussion of the electrostatic field we showed, by means of Gauss' theorem, that the field $\boldsymbol{E}$ caused by any assembly of charges acting in accordance with the inverse square law, must be a flux vector. By exactly similar arguments it follows that the magnetic field $\boldsymbol{H}$ and the flux density $\boldsymbol{B}$ produced by a magnetic shell in free space are both flux vectors. Combining this result with the equivalence of magnetic shell and current-carrying circuit, we can now assert that, in free space, $\boldsymbol{B}$ and $\boldsymbol{H}$ produced

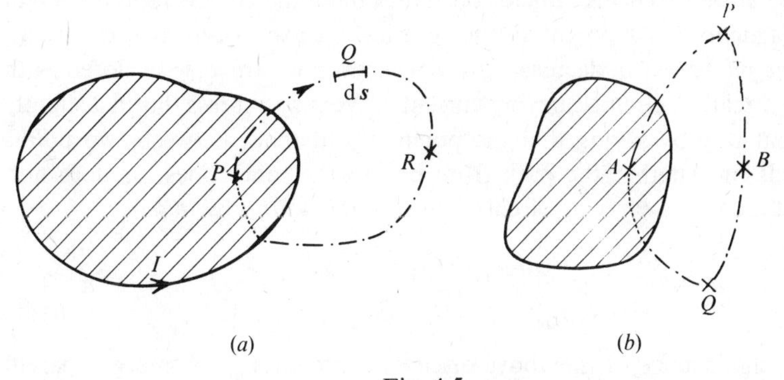


(*a*) (*b*)

Fig. 4.5

by any distribution of currents will always be flux vectors. This is the property of $\boldsymbol{B}$ which we assumed in §4.1.4.

### 4.1.10 The properties of the magnetic scalar potential $U$: Ampère's circuital law

So far, we have defined the magnetic scalar potential $U$ by (4.14),

$$\boldsymbol{H} = -\frac{\partial U}{\partial n}\boldsymbol{n} = -\operatorname{grad} U \tag{4.44}$$

We must now investigate the properties of $U$.

In fig. 4.5(*a*) let current $I$ be flowing in the direction of the arrow in a circuit which, for the moment, we assume to lie in a plane. $PQR$ represents any closed path which passes through the circuit and we wish to consider changes in $U$ and $\boldsymbol{H}$ as we proceed round this path in the direction of the arrow. We start from $P$, the point in which our path intersects the plane of the circuit. At this point the solid angle subtended by the circuit is $2\pi$. Let the magnetic potential there be $U_0$. As we move away from $P$, the solid angle decreases, and so also does $U$ in agreement with the fact (4.14) that a decrease in $U$ occurs for a displacement in the positive direction of $\boldsymbol{H}$. When we reach $R$, the second point in which our path intersects the plane of the circuit, the solid angle has been reduced to zero. As we pass beyond $R$ the solid angle becomes negative, since it is now the opposite surface of the area bounded by the circuit, which subtends the angle. Finally, as we move from $R$ to $P$, the angle increases in negative value until, at $P$, it is equal to $-2\pi$. Thus, in traversing the complete path, we have caused the solid angle to decrease by $4\pi$. In consequence, the potential at $P$ will have decreased to $U_1$ where, by (4.42),

$$U_0 - U_1 = I \tag{4.45}$$

We therefore reach the conclusion that, unlike the electrostatic potential, the magnetic scalar potential does not have a fixed value at every point, since it increases or decreases by $I$ whenever we traverse a closed path which threads a circuit-carrying current $I$. We may express this in a slightly different way by saying that the potential difference between two points depends on whether the path from one to the other does, or does not, thread a current-carrying circuit. Thus, in fig. 4.5(*b*), by (4.45),

$$(U_P - U_Q)_{PBQ} + (U_Q - U_P)_{QAP} = I$$

or

$$(U_P - U_Q)_{PBQ} - (U_P - U_Q)_{PAQ} = I \tag{4.46}$$

The significance of the above discussion becomes much more apparent when we state the results in terms of the magnetic field strength $\boldsymbol{H}$ rather than the scalar potential $U$. Reverting to fig. 4.5(*a*), if we consider an element $\mathrm{d}\boldsymbol{s}$ of the path at $Q$ and if $\boldsymbol{H}\cdot\mathrm{d}\boldsymbol{s}$ is the component of $\boldsymbol{H}$ along $\mathrm{d}\boldsymbol{s}$ at this point, the change $\mathrm{d}U$ in the magnetic potential as we traverse $\mathrm{d}\boldsymbol{s}$ is, by (4.14)

$$\mathrm{d}U = -\boldsymbol{H}\cdot\mathrm{d}\boldsymbol{s} \tag{4.47}$$

so we may write

$$\oint_s \mathrm{d}U = -\oint_s \boldsymbol{H}\cdot\mathrm{d}\boldsymbol{s} \tag{4.48}$$

However, from (4.45),

$$\oint_s \mathrm{d}U = -I \tag{4.49}$$

so that, finally

$$\oint_s \boldsymbol{H}\cdot\mathrm{d}\boldsymbol{s} = I \tag{4.50}$$

This is a result of the greatest importance; it is often known as Ampère's circuital law, though he did not state it explicitly in this form. In words, we may say that for any closed path in a magnetic field, the line integral of $\boldsymbol{H}$ round the path is equal to the current linked by the path. If the path is linked with more than one current-carrying circuit, the superposition theorem (§4.1.2) tells us that the line integral of $\boldsymbol{H}$ will be equal to the algebraic sum of the currents linked.

If we compare (4.50) with the corresponding equation (2.25) for the electrostatic field, we see that it is only in the special case when the path does not link any current, that we can write

$$\oint_s \boldsymbol{H}\cdot\mathrm{d}\boldsymbol{s} = 0$$

In general, the magnetic field is not conservative and the magnetic scalar potential does not have a single unique value. Nevertheless it is always valid to write

$$\boldsymbol{H} = -\operatorname{grad} U$$

since this equation relates only to the change in $U$ during an infinitesimal displacement.

## 4.2 The calculation of magnetic field strengths

### 4.2.1 Methods

The theory established in the foregoing sections provides three separate methods by which the magnetic field strength $\boldsymbol{H}$, and therefore the flux density $\boldsymbol{B}$, can be calculated. The most generally useful of these is expressed by (4.10).

$$\boldsymbol{H} = \frac{I}{4\pi}\oint_s \frac{\mathrm{d}\boldsymbol{l}\times\boldsymbol{r}_0}{r^2} \tag{4.51}$$

This equation can always be used though, if the circuit is not of simple geometrical form, the integration may have to be carried out numerically, with or without the aid of a computer.

As a second method we may calculate the magnetic potential from (4.42)

$$U = I\Omega/4\pi \tag{4.52}$$

and then derive $\boldsymbol{H}$ from (4.14)

$$\boldsymbol{H} = -\mathrm{grad}\, U \tag{4.53}$$

Finally, in a few special cases where, from conditions of symmetry, we can deduce the manner in which $\boldsymbol{H}$ varies round a particular path, we can make use of (4.50)

$$\oint_s \boldsymbol{H}\cdot\mathrm{d}\boldsymbol{s} = I \tag{4.54}$$

We shall now apply these methods to derive $\boldsymbol{H}$ for a few circuits of simple form.

### 4.2.2 The field on the axis of a circular coil

In fig. 4.6($a$) let the coil consist of $N$ turns of fine wire, each of radius $r$, carrying current $I$. We require the value of $\boldsymbol{H}$ at a point $P$ distant $x$ from the plane of the coil. This problem can readily be solved by the use of (4.51), but it is instructive to obtain the result from (4.52) and (4.53). For this purpose we need to calculate the solid angle which the surface bounded by the coil subtends at $P$. In fig. 4.6($b$) consider a sphere with centre at $P$ and of radius $R$ such that the coil lies on its surface. Thus

$$R^2 = r^2+x^2 \tag{4.55}$$

The area of the spherical surface lying between two planes at right angles

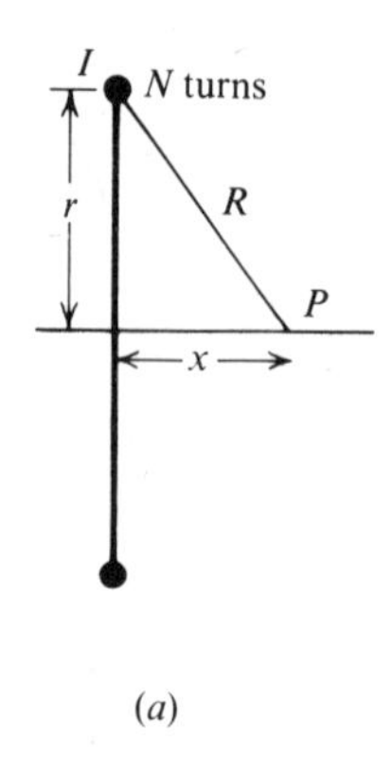


(a)

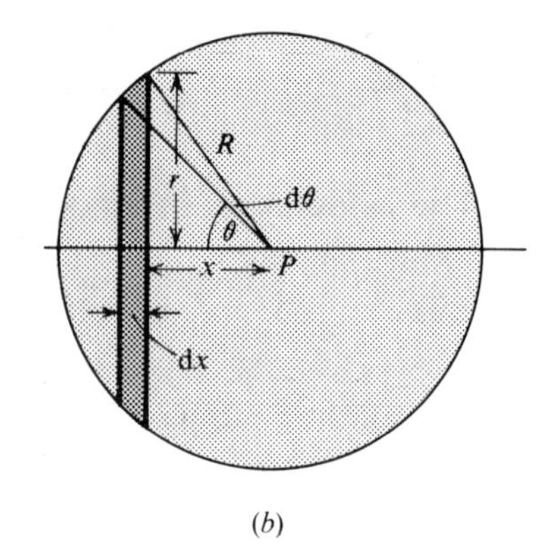


(b)

Fig. 4.6

to the axis of the coil and distant d$x$ apart is, with the notation of the figure,

$$R\,\mathrm{d}\theta\cdot 2\pi R\sin\theta = 2\pi R\,\mathrm{d}x \tag{4.56}$$

Reverting to fig. 4.6(*a*), it is clear that the area of the portion of the sphere, of radius $R$, lying to the left of the coil is

$$2\pi R(R-x)$$

and the solid angle $\Omega$ which this area subtends at $P$ is

$$\frac{2\pi R(R-x)}{R^2} = 2\pi\left(1-\frac{x}{R}\right) \tag{4.57}$$

For the magnetic potential $U$ we then have

$$U = \frac{NI}{2}\left(1-\frac{x}{R}\right) = \tfrac{1}{2}NI[1-x(r^2+x^2)^{-\frac{1}{2}}] \tag{4.58}$$

By symmetry, the magnetic field strength $H$ must lie along the axis of the coil, so

$$H = -\frac{\partial U}{\partial x} = \frac{NIr^2}{2(r^2+x^2)^{\frac{3}{2}}} \tag{4.59}$$

When $P$ lies in the plane of the coil, $x = 0$ and

$$H = NI/2r \tag{4.60}$$

The direction of $H$ along the axis follows from (4.51). If a right-handed screw lay along the axis and were twisted in the direction of current flow, the direction in which it moved axially would be the direction of $H$.

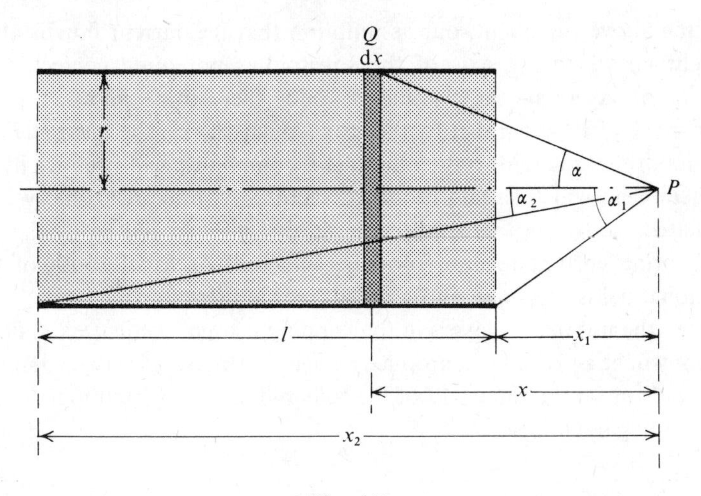


Fig. 4.7

### 4.2.3 The field on the axis of a solenoid

Let the solenoid have length $l$ and radius $r$, with $N$ turns carrying current $I$. We suppose the winding depth to be negligible in comparison with $r$ and the number of turns per unit length to be large, so that the direction of current flow is always very nearly at right angles to the axis. Then, from the symmetry of the system, the magnetic field $\boldsymbol{H}$ at any point of the axis will lie along the axis.

With the notation of fig. 4.7, let $\mathrm{d}H$ be the component of $H$ at point $P$, which is caused by an element $\mathrm{d}x$ of the solenoid situated at $Q$. The number of turns in $\mathrm{d}x$ is $N\mathrm{d}x/l$ so that, from (4.59),

$$\mathrm{d}H = \frac{NIr^2\mathrm{d}x}{2l(r^2+x^2)^{\frac{3}{2}}} \tag{4.61}$$

Hence
$$H = \frac{NI}{2l}\int_{x_1}^{x_2} \frac{r^2\mathrm{d}x}{(r^2+x^2)^{\frac{3}{2}}} \tag{4.62}$$

Using the substitution $x = r\cot\alpha$ this becomes

$$H = -\frac{NI}{2l}\int_{\alpha_1}^{\alpha_2} \sin\alpha\,\mathrm{d}\alpha$$

$$= \frac{NI}{2l}(\cos\alpha_2 - \cos\alpha_1) \tag{4.63}$$

The direction of $\boldsymbol{H}$ along the axis has already been discussed at the end of §4.2.2.

In the above treatment, our assumption that the current flow is always at right angles to the axis of the solenoid is not quite correct. If the number of layers in the winding is odd, there must necessarily be a component of flow parallel to the axis, which carries the current $I$ from one end of the solenoid to the other. This component will produce its own magnetic field which should be added vectorially to the field that we have calculated. Similarly, there will be an additional field caused by the leads which bring current to and from the solenoid. As a rule both of these additional fields are small and we shall ignore them.

When the number of layers in the winding is even, the nett axial flow of current will be zero and will produce no field. Moreover, the two connecting leads will run to the same end of the solenoid and, if twisted together, will produce negligible field.

### 4.2.4 The field near a finite straight wire carrying current

This is a hypothetical case, since a finite straight wire cannot by itself form a complete circuit. However, the result which we shall obtain is useful, since it can be used to derive $\boldsymbol{H}$ for any circuit consisting of a number of straight conductors.

Referring to fig. 4.8($a$) we require the value of $\boldsymbol{H}$ at point $P$, distant $r$ from the axis of the wire, whose radius is assumed to be negligible. The angles between the axis and lines drawn from $P$ to the ends of the wire are $\alpha_1$ and $\alpha_2$ respectively and $I$ is the current carried by the wire. $O$ is the foot of the perpendicular from $P$ on to the axis and we consider the component of $\boldsymbol{H}$ caused by current in an element d$x$ of the wire distant $x$ from $O$. The angle between the axis and the line from $P$ to this element is $\alpha$.

From (4.51) the component d$H$ caused by this element is normal to the plane of the paper and, with the direction of current flow indicated, is directed out of the paper. Its magnitude is given by

$$\mathrm{d}H = \frac{I\mathrm{d}x \sin \alpha}{4\pi PQ^2}$$

But

$$PQ = r \operatorname{cosec} \alpha$$

$$-x = r \cot \alpha$$

$$\mathrm{d}x = r \operatorname{cosec}^2 \alpha \, \mathrm{d}\alpha$$

so

$$\mathrm{d}H = \frac{I \sin \alpha \, \mathrm{d}\alpha}{4\pi r}$$

Since the components for all elements d$x$ are in the same direction, they

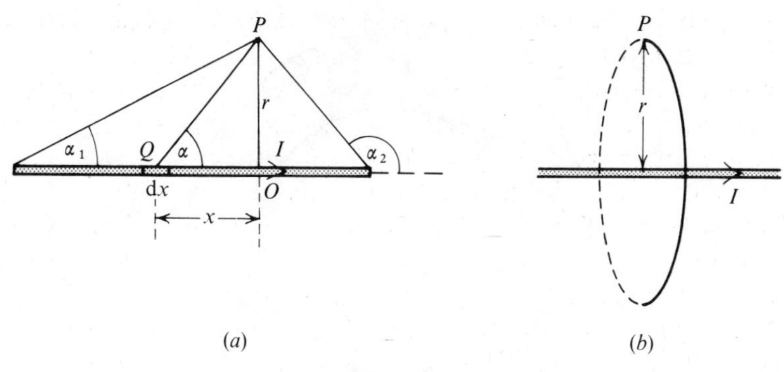


Fig. 4.8

will add numerically and

$$H = \frac{I}{4\pi r}\int_{\alpha_1}^{\alpha_2} \sin\alpha \, d\alpha$$

$$= \frac{I}{4\pi r}(\cos\alpha_1 - \cos\alpha_2) \quad (4.64)$$

For an infinitely long wire $\alpha_1 = 0$ and $\alpha_2 = \pi$, so

$$H = I/2\pi r \quad (4.65)$$

This last result can be obtained very readily from (4.54). With an infinitely long wire, the leads bringing current to and from the wire are so remote that they cannot affect $\boldsymbol{H}$ at the point $P$. If, therefore, we consider a circular path of radius $r$, with centre on the wire and passing through $P$ (fig. 4.8(*b*)) conditions of symmetry show that $\boldsymbol{H}$ will always lie along the circumference and will have the same value at all points on the path. Thus

$$\oint \boldsymbol{H}\cdot d\boldsymbol{s} = 2\pi r H = I$$

or

$$H = I/2\pi r \quad (4.66)$$

### 4.2.5 The field inside a uniformly wound toroid

In fig. 4.9 we suppose the toroid to be uniformly wound with a total of $N$ turns carrying current $I$. We also assume that each turn lies in a plane that is radial to the axis of the toroid. This is not quite true and, if the number of layers in the winding is odd, there will be a nett circumferential flow of current round the toroid, producing effects similar to those which we discussed in connection with the solenoid (§4.2.3). When the number of layers is even, these effects largely cancel and this is the case that we shall consider.

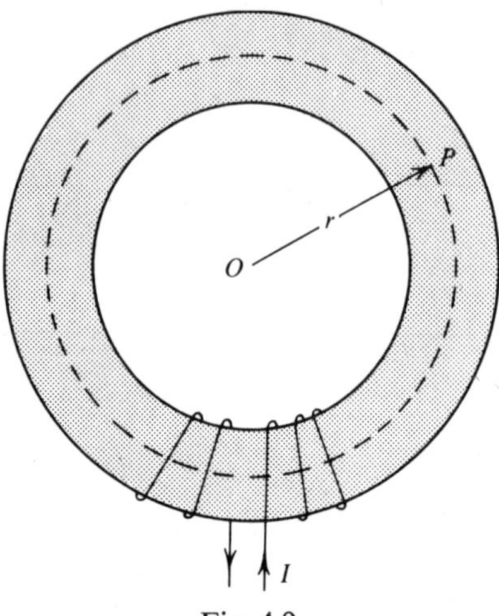


Fig. 4.9

We wish to find $\boldsymbol{H}$ at a point $P$, inside the toroid, where the distance from $P$ to the axis $O$ is $r$. From the symmetry of the system we may conclude that $\boldsymbol{H}$ is normal to the axis at all points inside the toroid. To see this we note that, for any element of current which could produce a radial component of $\boldsymbol{H}$ or a component parallel to the axis, there is a corresponding element producing opposite effects in these directions. We therefore consider a circular path of radius $r$, in a plane normal to the axis and with centre at $O$. $\boldsymbol{H}$ will be constant along this path so, by (4.54),

$$\oint_s \boldsymbol{H}\cdot \mathrm{d}\boldsymbol{s} = 2\pi r H = NI$$

or

$$H = NI/2\pi r \tag{4.67}$$

By similar reasoning we find that, for a point $P$ outside the toroid, $\boldsymbol{H}$ is zero. These results are independent of the shape and size of the cross-section of the toroid.

Since the value of $\boldsymbol{H}$ in the toroid depends on $r$, the field is not uniform. However, by keeping the radial dimension of the cross-section small in comparison with $r$, we can make the field as nearly uniform as we wish. Because of this property, a toroidal winding is of considerable importance in magnetic measurements.

## 4.3 Forces on charged particles moving in a magnetic field

### 4.3.1 The basic postulates

In the earlier part of this chapter we have been concerned with the description of the magnetic field in terms of the vectors $\boldsymbol{B}$ and $\boldsymbol{H}$ and with the calculation of these quantities. We must now turn our attention to the measurable physical effects which can be calculated from a knowledge of $\boldsymbol{B}$.

In chapter 3 we dealt with the force acting on a particle with charge $Q$ in an electrostatic field $\boldsymbol{E}$. If the particle is moving through a magnetic field with velocity $\boldsymbol{u}$ it experiences an additional force $\boldsymbol{F}$, which can be calculated according to the following rules:

(*a*) $\boldsymbol{F}$ has magnitude $QuB \sin\theta$, where $\theta$ is the angle between $\boldsymbol{B}$ and $\boldsymbol{u}$;
(*b*) it acts along a line which is normal to the plane containing $\boldsymbol{B}$ and $\boldsymbol{u}$;
(*c*) it is in the direction moved by a right-handed screw in twisting from $\boldsymbol{u}$ to $\boldsymbol{B}$.

Thus in vector notation we can write

$$\boldsymbol{F} = Q\boldsymbol{u}\times\boldsymbol{B} \text{ newtons} \tag{4.68}$$

If an electrostatic field $\boldsymbol{E}$, as well as the magnetic field $\boldsymbol{B}$, is present, the total force $\boldsymbol{F}'$ on the particle is given by

$$\boldsymbol{F}' = Q(\boldsymbol{E}+\boldsymbol{u}\times\boldsymbol{B}) \tag{4.69}$$

$\boldsymbol{F}'$ is known as the *Lorentz force* and this term is frequently applied to $\boldsymbol{F}$ also.

It is of interest to note that (4.69) can be deduced from electrostatic theory, by way of the theory of relativity, but we shall not pursue this matter. We shall adopt the equation as a basic postulate which is justified by experiment. It has, in fact, been verified with great accuracy by experiments on electron beams and on beams of other charged particles.

We have already assumed that an electric current is the result of a flow of charged particles (§2.1.1). We now make the additional assumption that, when a current-carrying conductor is situated in a magnetic field, the particles which are carrying the current will experience forces in accordance with (4.68). Since, as a rule, the particles cannot escape from the conductor, the forces acting on them will be transferred to the conductor.

Consider a circuit-carrying current $I$ and let the cross-section of the conductor be uniform and small, so that $\boldsymbol{B}$ may be taken to be constant over any cross-section. Let $I$ be the result of charge $Q$ per unit length of conductor, moving with average velocity $u$. Then an element $\mathrm{d}\boldsymbol{s}$ of the circuit which is situated in a magnetic field of flux density $\boldsymbol{B}$, will experience a force $\mathrm{d}\boldsymbol{F}$ given by

$$\mathrm{d}\boldsymbol{F} = Qu\mathrm{d}\boldsymbol{s}\times\boldsymbol{B} = I\mathrm{d}\boldsymbol{s}\times\boldsymbol{B} \tag{4.70}$$

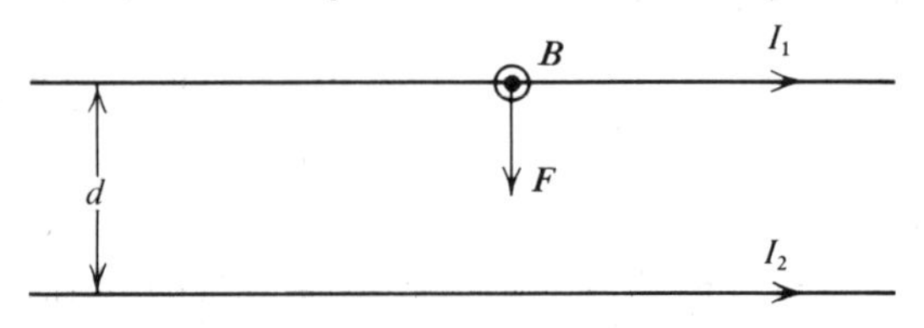


Fig. 4.10

For the total force $\boldsymbol{F}$ on the circuit we have

$$\boldsymbol{F} = \oint_s I \mathrm{d}\boldsymbol{s} \times \boldsymbol{B} \tag{4.71}$$

If the cross-section of the conductor is neither small nor uniform, the total current must be divided into filaments and the total force on the conductor found by integration.

Equation (4.71) has also been verified with great accuracy in a large number of experiments and we adopt it as our second basic postulate.

We shall now use (4.71) to calculate forces in two cases of particular importance. An alternative method of deriving the force or couple on a circuit will be discussed later (§8.2.3).

### 4.3.2 The permeability of free space

The definition of the ampere (§2.1.1) is based on the force between two parallel, infinitely long, current-carrying wires, of negligible cross-section. We are now in a position to calculate this force.

In fig. 4.10, let the two conductors carry currents $I_1$ and $I_2$ respectively and be separated by distance $d$. From (4.65), the magnitude of the flux density, at any point of $I_1$, produced by $I_2$, is given by

$$B = \mu_0 I_2 / 2\pi d$$

$\boldsymbol{B}$ is normal to the plane of the diagram and is directed out of the page. Thus the force $\boldsymbol{F}$ per unit length of the system has the direction shown and from (4.70) is of magnitude

$$F = \mu_0 I_1 I_2 / 2\pi d \text{ newtons metre}^{-1} \tag{4.72}$$

From the definition of the ampere, if

$$I_1 = I_2 = 1 \text{ A} \quad \text{and} \quad d = 1 \text{ m}$$

$$F = 2 \times 10^{-7} \text{ newtons metre}^{-1}$$

Thus it follows from this definition and from (4.72) that $\mu_0$ must have the *exact* value

$$\mu_0 = 4\pi \times 10^{-7} \text{ H m}^{-1} \tag{4.73}$$

The units used for $\mu_0$ have already been explained (§4.1.5).

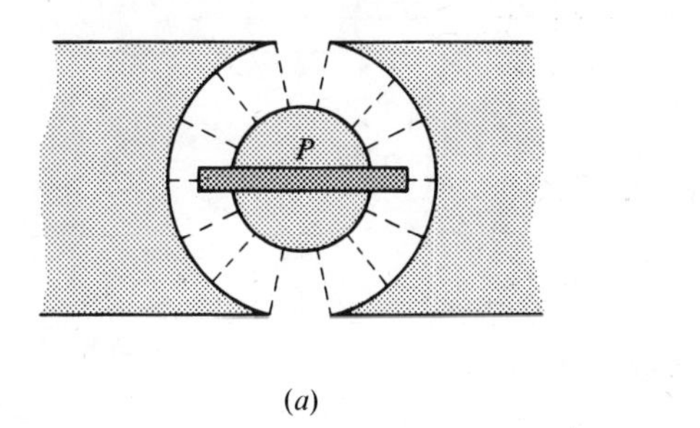


(*a*)

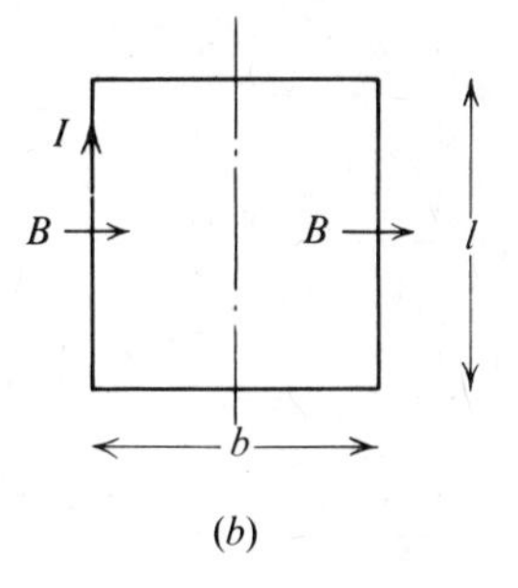


(*b*)

Fig. 4.11

### 4.3.3 The couple exerted in a moving-coil meter

In a moving-coil meter or galvanometer the coil which, for the moment, we suppose to be rectangular, is suspended between the poles of a permanent magnet, shaped as shown in fig. 4.11(*a*). A cylinder of soft iron $P$ is mounted inside, but not touching, the coil, so that the magnetic flux density $\boldsymbol{B}$ (shown as broken lines in fig. 4.11(*a*)) is radial and remains constant when the coil is deflected about its axis perpendicular to the plane of the diagram. Current $I$ is led to and from the coil through suspensions (not shown in the diagram) which also exert a restoring couple when the coil is deflected. Let the coil have $N$ turns, with the dimensions shown in fig. 4.11(*b*). The sides of length $b$ lie outside the magnetic field, so no force acts on them. The sides of length $l$ lie in a field of flux density $B$, at right angles to these sides, so each will experience a force $NIBl$, normal to the plane of the coil. The forces on the two sides are in opposite directions, so they form a couple of magnitude

$$T = NIBlb = NIBA \tag{4.74}$$

where $A$ is the area of the coil.

It is left as an exercise for the reader to show that the torque $T$ acting on the coil is equal to $NIBA$, even if the shape of the coil is not rectangular.

## 4.4 Electromagnetic induction

### 4.4.1 The motion of a conductor through a magnetic field

We have seen that, when a charge $Q$ moves through a magnetic field of flux density $\boldsymbol{B}$ with velocity $\boldsymbol{u}$, it experiences a force $\boldsymbol{F}$ given by (4.68)

$$\boldsymbol{F} = Q\boldsymbol{u}\times\boldsymbol{B} \tag{4.75}$$

Hitherto we have supposed the relative notion of charge and field to have been brought about by the flow of current in a stationary conductor. There

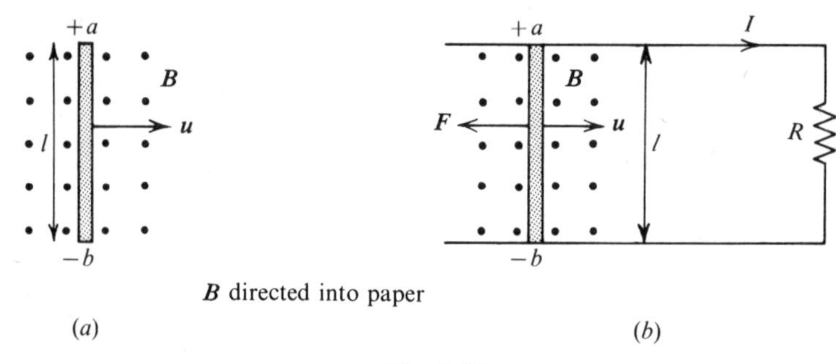


Fig. 4.12

is, however, the alternative possibility that the charge should remain in a fixed position relative to the conductor and that the whole conductor should move bodily through the field.

In fig. 4.12(*a*) let there be a magnetic flux density $\boldsymbol{B}$ (represented by dots) normal to the plane of the diagram and directed into the paper. A straight conductor *ab* of length $l$ is moving at right angles to the magnetic field with velocity $\boldsymbol{u}$. In accordance with (4.75) we should expect the mobile charges inside the conductor, whether positive or negative, to experience forces urging positive charge towards *a* and negative charge towards *b*. This redistribution of charge will set up an electric field $\boldsymbol{E}$ within the conductor, opposing further motion of the charges, and equilibrium will be obtained when

$$-\boldsymbol{E}Q = Q\boldsymbol{u}\times\boldsymbol{B}$$

or

$$-\boldsymbol{E} = \boldsymbol{u}\times\boldsymbol{B} \tag{4.76}$$

giving rise to a potential difference $V$ between *a* and *b*, where

$$V = -\boldsymbol{E}\cdot\boldsymbol{l} = \boldsymbol{l}\cdot(\boldsymbol{u}\times\boldsymbol{B}) \tag{4.77}$$

*a* being positive with respect to *b*.

We cannot test this conclusion experimentally without allowing the conductor to form part of a complete circuit. In principle we can do this, fig. 4.12(*b*), by supposing the rod *ab* to slide along two parallel straight conductors whose far ends are joined by a conductor of resistance $R$. To simplify matters we suppose $R$ to be the total resistance of the complete circuit. The potential difference $V$ (4.77) will cause current

$$I = [\boldsymbol{l}\cdot(\boldsymbol{u}\times\boldsymbol{B})]/R \tag{4.78}$$

to flow round the circuit and energy will be dissipated as heat at a rate of

$$W = I^2R = \{\boldsymbol{l}\cdot(\boldsymbol{u}\times\boldsymbol{B})\}^2R \text{ watts} \tag{4.79}$$

To produce this heat, mechanical work must have been done and the source of the work lies in the force acting on the conducting rod. Since the rod carries current $I$ and is situated in a magnetic field, it will experience a force

$$\boldsymbol{F} = I\boldsymbol{l}\times\boldsymbol{B} \tag{4.80}$$

and work will be done at a rate

$$W' = I(\boldsymbol{l}\times\boldsymbol{B})\cdot\boldsymbol{u} \tag{4.81}$$

In the simple arrangement represented by fig. 4.12(*b*), $\boldsymbol{l}$, $\boldsymbol{B}$ and $\boldsymbol{u}$ are all at right angles to each other, while $\boldsymbol{F}$ is normal to $\boldsymbol{l}$ and $\boldsymbol{B}$ and is in the opposite direction to $\boldsymbol{u}$. Thus (4.79) becomes

$$W = l^2u^2B^2/R \tag{4.82}$$

and from (4.78) and (4.81)

$$W' = (luB/R)(luB) = l^2u^2B^2/R \tag{4.83}$$

Thus the mechanical work done in moving the rod with velocity $u$ is exactly equal to the energy dissipated as heat in the resistance $R$. In general, the flow of current $I$ will cause forces on other parts of the circuit as well as on the rod. However, since these other parts do not move, no work is done on or by them.

### 4.4.2 Electromotive force and electromagnetic induction

Our discussion of the system of fig. 4.12(*b*) has shown that magnetic as well as electrostatic forces can act on the charges within a conducting circuit and that it is the former which are directly responsible for producing a potential difference between $a$ and $b$. In the portion of the circuit from $a$ to $b$ via $R$, any magnetic force on the moving charge will be at right angles to the direction of motion and so will do no work. Thus the work done in taking unit charge from $a$ to $b$ via $R$ is work done by electrostatic forces and is equal to the potential difference between $a$ and $b$.

This, however, is no longer true for the return path from $b$ to $a$ along the rod. Here, as we have seen, the force which the electrostatic field exerts on a charge is exactly balanced by the magnetic force resulting from the motion of the charge through the magnetic field. Thus no work is done when the charge is taken from $b$ to $a$ along the rod.

Considering the complete circuit, we see that, when unit charge is taken from $a$ to $b$ via $R$ and then back to $a$ along the rod, a nett quantity of work is done, which is equal to the potential difference between $a$ and $b$. It is no longer true, as it was in a purely electrostatic field, to say that the total potential change round any closed path is zero.

We must expect this situation to arise in any circuit where forces other than purely electrostatic forces are acting. In addition to the case that we have been considering, other examples are circuits connected to a battery, where chemical energy causes the flow of current, and circuits in which thermocouples convert thermal to electrical energy. For such circuits we need a new quantity to replace potential difference and to take account of *all* forces acting on the charges in the circuit. We therefore *define* the *electromotive force* (e.m.f.) acting in a circuit to be the work done when unit charge is taken once round the circuit. The term potential difference is best restricted to the electrostatic field or to portions of a circuit where only electrostatic forces are acting. Thus one might speak of the potential difference between the terminals of a battery or across any component of an external circuit connected to the battery. It would be unwise to speak of potential differences within the battery itself without giving very careful consideration to sources of chemical energy. For a complete circuit including the battery it would be appropriate to speak of the electromotive force. The symbol $E$ is commonly used for e.m.f.; we have also used this symbol for the magnitude of the vector electrostatic field strength $\boldsymbol{E}$. In general, it is obvious from the context which quantity is under consideration. The unit of $E$, as of potential difference, is the volt.

In the light of the above discussion we revert to the system of fig. 4.12(*b*) and write (4.77) in the form

$$\text{e.m.f.} = E = \boldsymbol{l}\cdot(\boldsymbol{u}\times\boldsymbol{B}) \text{ volt} \tag{4.84}$$

or, since $\boldsymbol{l}$, $\boldsymbol{u}$ and $\boldsymbol{B}$ are all at right angles,

$$E = luB \tag{4.85}$$

We note that $lu\mathrm{d}t$ is the area swept out by the rod in time $\mathrm{d}t$ and that $luB\mathrm{d}t$ is the decrease, during this interval, in the flux $\phi$ linked with the circuit (cf. §4.1.4, (4.7)). Thus we write

$$-\mathrm{d}\phi = luB\mathrm{d}t$$

and, from (4.85)

$$E = -\frac{\mathrm{d}\phi}{\mathrm{d}t} \tag{4.86}$$

$E$ is referred to as the e.m.f. *induced* in the circuit by the motion of the rod through the magnetic field and the phenomenon is termed *electromagnetic induction.*

### 4.4.3 The flux-cutting rule; Lenz's law

In the system of fig. 4.12(*b*), as the rod *ab* moves it passes through magnetic flux and may be said to cut this flux. Moreover, the amount of flux cut in time $dt$ is $luB\,dt$ so that, from (4.85), we see that *the induced e.m.f. is equal to the rate at which flux is being cut*. We shall shortly generalize this result, which is commonly known as the *flux-cutting rule*.

We have seen that the current which flows in the circuit of fig. 4.12(*b*) is in such a direction as to exert a force opposing the motion of the rod *ab*. Since power is dissipated in the circuit, this is a necessary consequence of the conservation of energy. The flow of current also opposes the change brought about by the motion of the rod in another sense. In the figure, the flux density $\boldsymbol{B}$ was directed into the paper and the motion of the rod causes a reduction in the amount of flux linked with the circuit. However, the resulting induced current itself produces flux and, within the circuit, this flux is directed into the paper. Thus the induced current opposes the change in the quantity of flux linked with the circuit. The opposition to the change on both of these counts is an example, for this particular system, of a law first stated by Lenz, as follows. *When the magnetic flux linked with a circuit changes, the direction of the induced e.m.f. is such that any current produced by it tends to oppose the change of flux.*

### 4.4.4 Generalization of the above results

In order to concentrate attention on the physical principles involved, we have so far considered only the very simple system of fig. 4.12(*b*), but it is clearly desirable to generalize the results for a circuit of any form moving in a magnetic field of any configuration.

In fig. 4.13, an element $d\boldsymbol{s}$ of a circuit is moving with velocity $\boldsymbol{u}$ through a magnetic field of flux density $\boldsymbol{B}$. The parallelepiped of which these three vectors form edges, has been completed to give a better indication of the directions. The force on a positive charge in $d\boldsymbol{s}$ is in the direction of the vector $\boldsymbol{u}\times\boldsymbol{B}$ and this will not usually coincide with the direction of $d\boldsymbol{s}$ itself. We therefore write (4.77) in the form

$$dE = d\boldsymbol{s}\cdot(\boldsymbol{u}\times\boldsymbol{B}) \tag{4.87}$$

where $dE$ is the component of e.m.f. induced in $d\boldsymbol{s}$. We also have the vector equation

$$d\boldsymbol{s}\cdot(\boldsymbol{u}\times\boldsymbol{B}) = \boldsymbol{B}\cdot(d\boldsymbol{s}\times\boldsymbol{u}) \tag{4.88}$$

since each of these quantities is equal to the volume of the parallelepiped. But $d\boldsymbol{s}\times\boldsymbol{u}$ is a vector whose magnitude is equal to the area swept out by $d\boldsymbol{s}$ in unit time and whose direction is at right angles to this area. Thus $\boldsymbol{B}\cdot(d\boldsymbol{s}\times\boldsymbol{u})$ is the rate at which $d\boldsymbol{s}$ is cutting flux and, by (4.87) and (4.88), this is equal to $dE$. For the complete circuit we have

$$E = \oint_s d\boldsymbol{s}\cdot(\boldsymbol{u}\times\boldsymbol{B}) \tag{4.89}$$

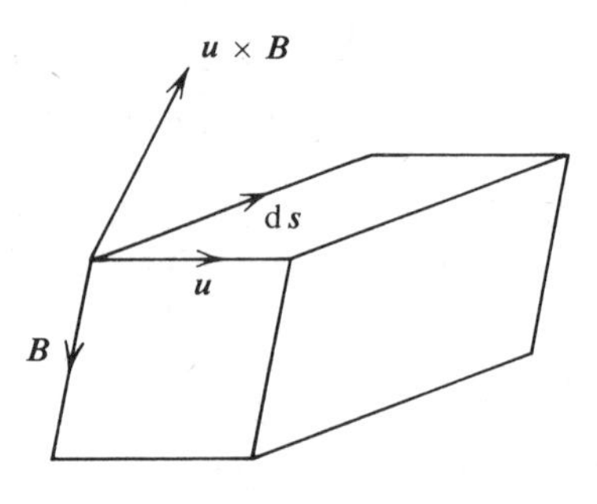


Fig. 4.13

Also, since the rate at which the flux $\phi$ linked with the circuit is changing is equal to the algebraic sum of the rates at which the elements are cutting flux, we may write, as in (4.86)

$$E = -\frac{d\phi}{dt}$$

The negative sign in this equation indicates that, if the directions of the e.m.f. $E$ round the circuit and the flux $\phi$ through the circuit are related in the same way as the twist and forward motion of a right-handed screw, then positive $E$ is caused by a decrease in $\phi$. It is also an expression of Lenz's law, which must be true for the complete circuit, since it holds for each element $d\boldsymbol{s}$ of the circuit.

### 4.4.5 Time-varying fields; the basic postulate

So far, our discussion of electromagnetic induction has been based on the assumption that the Lorentz equation (4.69) for the force on a charged particle moving through a magnetic field is valid for the mobile charges in conductors. While this is plausible, it really needs experimental confirmation. Moreover, the flux $\phi$ linked with a circuit can be changed by methods to which the Lorentz equation is not obviously applicable. For example, the circuit can remain at rest, while a permanent magnet or a current-carrying coil is moved in its vicinity. Again, with the circuit at rest, the current through an adjacent coil can be caused to vary with time. The early experiments of Faraday, and the innumerable precise measurements of induced e.m.f. that have been made since his time, show that in all these cases the following postulates are valid.

(i) Whenever the flux linked with a circuit changes, there is induced in the circuit an e.m.f. whose magnitude is equal to the time rate of variation of the flux.

(ii) The direction of the induced e.m.f. is such that any current caused by it tends to oppose the change of flux which produced it (Lenz's law).

We adopt these postulates as experimental facts. They are summarized in the equation

$$E = -\frac{d\phi}{dt} \tag{4.90}$$

In the general case, the flux linked with a circuit may alter because

(*a*) the value of $\boldsymbol{B}$ at any point of a surface $S$ bounded by the circuit is changing with time. This gives rise to a rate of increase of linked flux at

$$\frac{d\phi}{dt} = \int_S \frac{\partial \boldsymbol{B}}{\partial t} \cdot \boldsymbol{n}\, dS$$

(*b*) each element $d\boldsymbol{s}$ of the circuit is moving with velocity $\boldsymbol{u}$ through the field. The corresponding rate of increase of linked flux is

$$\frac{d\phi}{dt} = -\oint_s (\boldsymbol{u} \times \boldsymbol{B}) \cdot d\boldsymbol{s}$$

Taking both of these effects into account, we may write for the e.m.f. induced in the circuit

$$\text{e.m.f.} = E = -\frac{d\phi}{dt} = -\frac{d}{dt}\int_S \boldsymbol{B} \cdot \boldsymbol{n}\, dS$$

$$= \oint_s (\boldsymbol{u} \times \boldsymbol{B}) \cdot d\boldsymbol{s} - \int_S \frac{\partial \boldsymbol{B}}{\partial t} \cdot \boldsymbol{n}\, dS \tag{4.91}$$

### 4.4.6 The meaning of flux linkage

Hitherto we have considered flux linked with a circuit consisting of a single loop of wire which, by implication, has had negligible area of cross-section. Practical circuits may be more complicated.

It often happens that a coil contains $N$ turns of fine wire, which are bunched together so that the turns are very nearly coincident in space. In a case of this kind it is easy to see that the effective flux linked with the coil is approximately $N\phi_1$, where $\phi_1$ would be the flux linked with a single turn. We shall refer to $N\phi_1$ as the *flux linkage*.

In the case of a more complicated coil we may have the situation depicted in fig. 4.14(*a*), where two 'tubes' of magnetic flux are indicated by dashed lines. It is clear that the flux $d\phi_1$ is linked with one turn of the coil, while $d\phi_2$ is linked with three turns. The total flux linkage will be an intricate summation of terms, each term being the product of the flux within a small tube and the number of turns with which that tube is linked. We can

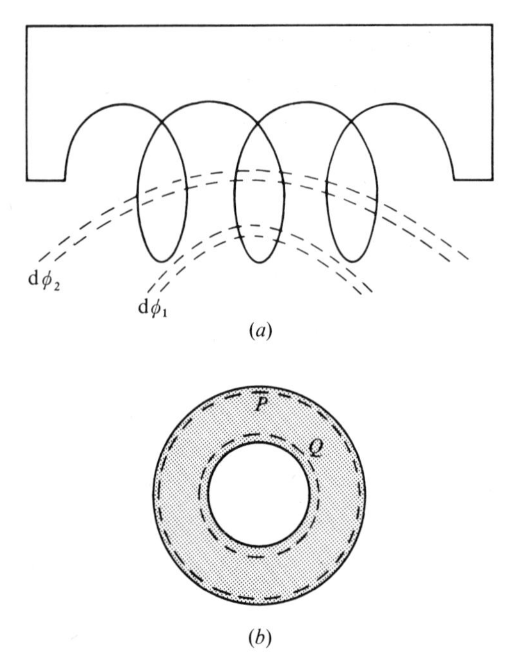


Fig. 4.14

express this relation formally as in (4.7) by writing

$$\text{flux linkage} = \phi = \int_S \boldsymbol{B}\cdot\boldsymbol{n}\,\mathrm{d}S \tag{4.92}$$

where $S$ is any surface bounded by the circuit and $\boldsymbol{n}$ is the unit normal to an element $\mathrm{d}S$ of this surface. Unless the circuit is a very simple one, it is usually difficult to visualize the form of $S$ and the evaluation of the integral in (4.92) may be a formidable undertaking. The important point is that (4.92) provides an unambiguous definition of what is meant by flux linkage.

If the cross-section of the conductor which forms the circuit is not negligible, further complications arise. Thus, in fig. 4.14(*b*), let a conducting ring of finite cross-section be situated in a magnetic field directed at right angles to the paper. The flux linkage with path $P$ is greater than that with path $Q$ and, if the field changes, the e.m.fs round these paths will be different. In addition to the main current flowing round the ring subsidiary currents will flow in the conductor and we generally refer to such currents as *eddy currents*. As a rule their calculation is a matter of considerable difficulty, but we shall later consider some important problems of this kind (§10.3).

### 4.4.7 Circuits with sliding contacts

We have seen that an e.m.f. may be induced in a circuit when the circuit moves through a constant magnetic field and that the e.m.f. is then given by (4.89)

$$E = \oint_S \mathrm{d}\boldsymbol{s}\cdot(\boldsymbol{u}\times\boldsymbol{B}) \tag{4.93}$$

It has also been shown that, for the circuits considered so far, (4.93) is equivalent to

$$E = -\frac{\mathrm{d}\phi}{\mathrm{d}t} \tag{4.94}$$

where $\phi$ is the flux linkage of the circuit, and that (4.94) is also valid when the circuit is at rest and the magnetic field is changing with time.

When the circuit under consideration contains one or more sliding contacts so that part of the circuit can move while the remainder is stationary, the interpretation of the above equations needs further thought. As an example of a problem of this type we take the *Faraday disc* illustrated in fig. 4.15. A circular conducting disc of radius $r_0$ rotates with angular velocity $\omega$ about a vertical conducting spindle, whose radius is small enough to be neglected. At right angles to the plane of the disc there is a magnetic field of flux density $\boldsymbol{B}$. Sliding contacts press on the disc at $Q$ and on the spindle at $T$ and the circuit is completed through a voltmeter $V$, which measures the induced e.m.f.

If we take the portion of the circuit which the disc contributes to lie along $PQ$, it is clear that the flux linked with the circuit does not change as the disc rotates. Thus, according to (4.94), no e.m.f. should be induced. On the other hand, a mobile charge in the disc situated on the line $PQ$ at distance $r$ from the axis, is moving with velocity $r\omega$ at right angles to $\boldsymbol{B}$ and will experience a force along $PQ$. Then, according to (4.93) there will be an induced e.m.f. given by

$$E = \int_0^{r_0} r\omega B\,\mathrm{d}r = \tfrac{1}{2}\omega B r_0^2 \tag{4.95}$$

and this is the e.m.f. which is found by experiment.

We can resolve this apparent contradiction if we define our circuit in a slightly different way. During an infinitesimal time interval $\mathrm{d}t$ a point on the rim of the disc, which was initially at $Q$, has moved to $S$, where $PS$ makes angle $\mathrm{d}\theta = \omega\,\mathrm{d}t$ with $PQ$. If we now take $PQT$ as the original circuit and $PSQT$ as the changed circuit after interval $\mathrm{d}t$, we see that the flux linked with the circuit has changed by the flux $\mathrm{d}\phi$ passing through the sector $PSQ$. But

$$\mathrm{d}\phi = -\tfrac{1}{2}r_0^2 B\omega\,\mathrm{d}t$$

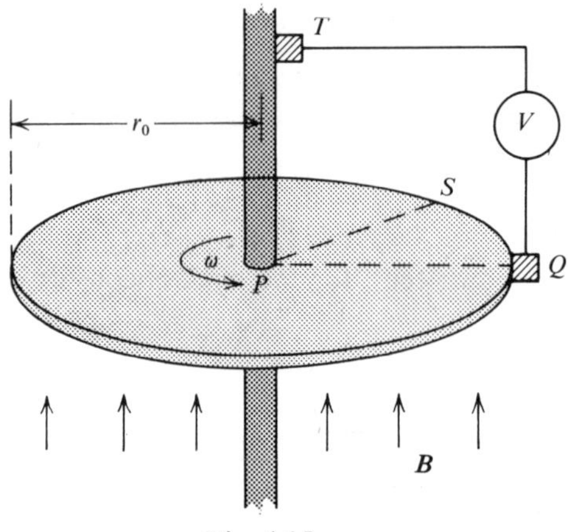


Fig. 4.15

so that, by (4.94),
$$E = -\frac{d\phi}{dt} = \tfrac{1}{2}\omega B r_0^2 \tag{4.96}$$
is agreement with (4.95).

We thus arrive at the following rule. *Equation* (4.94) *always gives the induced e.m.f. correctly, provided the flux linkage is evaluated for a circuit so chosen that at no point are particles of the conductor moving across the circuit.* It is to be noted that, in applying this rule, we must consider infinitesimally small changes in the circuit, since ***B*** will not necessarily be uniform.

## 4.5 Mutual inductance and self-inductance

### 4.5.1 Definition of mutual inductance

Fig. 4.16(*a*) represents a situation which often occurs in practical work. A generator $G$ causes current $I_1$ to flow round circuit 1 and a magnetic field is thereby set up. Some of the magnetic flux is linked with an adjacent circuit 2. If $I_1$ varies, so also will the flux linkage with 2 and an e.m.f. will be induced in 2.

At every point of the magnetic field the flux density ***B*** will be proportional to $I_1$, so the total flux linkage with 2 will also be proportional to $I_1$. We may therefore write for this flux linkage $\phi_{21}$
$$\phi_{21} = M_{21} I_1 \tag{4.97}$$
where $M_{21}$ is a constant known as the *mutual inductance* of the two circuits, when current flows in 1. It is the flux linkage with circuit 2 when unit current flows in circuit 1. If we differentiate both sides of (4.97) with

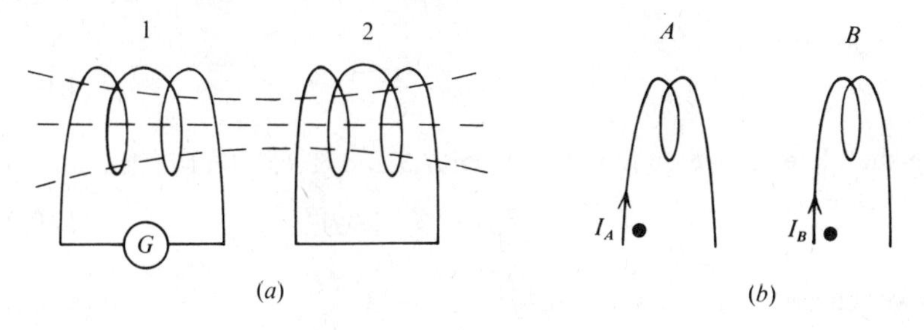


Fig. 4.16

respect to time, to find the e.m.f. $E_2$ that is induced in 2, we have

$$E_2 = -\frac{d\phi_{21}}{dt} = -M_{21}\frac{dI_1}{dt} \tag{4.98}$$

The negative signs in the above equation have no meaning unless we have a consistent convention for the positive directions of current in the two circuits and for the positive directions of the windings of the coils. If the circuits consist of parallel loops of wire with a common axis, there is no difficulty in establishing such a convention in accordance with the laws of vector algebra. However, if the circuits have distributed windings in random directions, the matter is less straightforward. In electrical equipment where the direction of the induced e.m.f. is important, it is common practice to mark *corresponding ends* of the two windings with a dot, as in fig. 4.16(*b*). This means that positive current $I_A$ entering the dot-marked terminal of circuit $A$, produces in circuit $B$ a flux in the same direction as would positive current $I_B$ entering the dot-marked terminal of circuit $B$. It follows that a positive rate of change of $I_A$ would cause the potential of the dot-marked terminal of $B$ to rise above that of the unmarked terminal. If an external circuit is connected to these terminals, positive current will flow outward from the dot-marked terminal, thus opposing the change of flux linked with $B$, in accordance with Lenz's law. In view of the above discussion we shall re-write (4.98) as

$$E_2 = M_{21}\frac{dI_1}{dt} \tag{4.99}$$

where it is to be understood that $M_{21}$ may be positive or negative, depending on the directions of winding of the two circuits and the conventions adopted for the positive directions of current flow.

So far we have assumed that, in fig. 4.16(*a*), current was varying in 1 and the e.m.f. was induced in 2, but this situation could be reversed. We should then define $M_{12}$ as the flux linked with 1 when unit current flows in 2, and write

$$E_1 = M_{12}\frac{dI_2}{dt} \tag{4.100}$$

We shall later prove quite generally (§6.7.3 and §8.1.6) that for circuits in free space

$$M_{12} = M_{21} = M \quad \text{(say)} \tag{4.101}$$

so we may write

$$E_1 = M\frac{dI_2}{dt}, \quad E_2 = M\frac{dI_1}{dt} \tag{4.102}$$

The unit of mutual inductance is the *henry* (symbol H). A pair of coils has a mutual inductance of one henry if a current of one ampere in one coil produces a flux linkage of one weber in the other. A pair of coils constructed to have a specified value of mutual inductance forms an important component in electrical measurements. It is known as a *mutual inductor*.

In the above discussion we have tacitly assumed that the conductors forming the two circuits have negligible cross-section and, for many purposes, this is a sufficiently good approximation. If the circuit 1 carrying the current $I_1$ has large cross-section, it is possible to consider the total current as being made up of a bundle of current filaments. The flux linkage with 2 can be calculated for each filament and the total flux linkage determined by summation. If the cross-section of circuit 2 is large, it is no longer accurate to speak of the e.m.f. induced in 2, since the e.m.f. will depend on which particular path round the circuit we choose. In general, the e.m.fs for different paths will not be the same and eddy currents will flow in the conductor. It is clear that any problem involving conductors with large cross-sections is likely to encounter formidable difficulties.

### 4.5.2 The mutual inductance of coaxial circular coils

There are very few pairs of circuits for which the mutual inductance can be calculated without great mathematical complication, but one such pair is illustrated in fig. 4.17. Two circular coils of fine wire have radii $r_1$ and $r_2$, and numbers of turns $N_1$ and $N_2$ respectively. The coils have a common axis and their planes are distant $x$ apart. The radius $r_2$ is small compared with $r_1$, so that the flux density over the whole of the small coil may be assumed to have a value equal to that on the axis.

For unit current in coil 1, the flux density at coil 2 is, by (4.59)

$$B = \mu_0 H = \frac{\mu_0 N_1 r_1^2}{2(r_1^2+x^2)^{\frac{3}{2}}}$$

and, with our approximation, is parallel to the axis. The mutual inductance

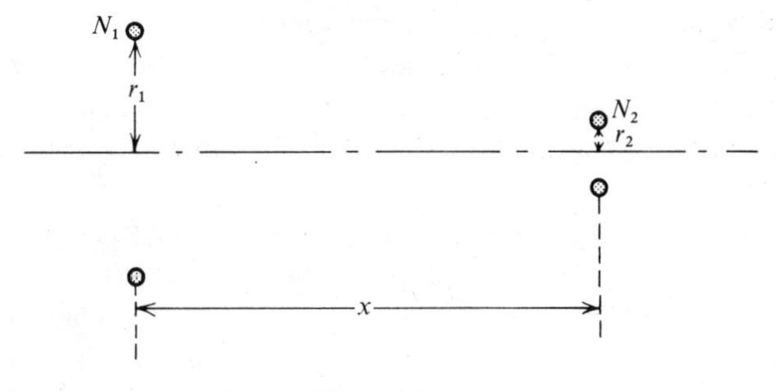


Fig. 4.17

is the flux linkage with coil 2 and is given by

$$M = B\pi r_2^2 N_2 = \frac{\pi\mu_0 N_1 N_2 r_1^2 r_2^2}{2(r_1^2+x^2)^{\frac{3}{2}}} \text{ henry} \tag{4.103}$$

If the two coils have comparable radii, (4.103) is no longer accurate and the calculation of $M$ becomes much more difficult. Comprehensive tables are available, from which $M$ can be found for all cases likely to arise in practice.

### 4.5.3 Self-inductance

When current $I$ flows in a circuit, a magnetic field will be established and the flux density at all points will be proportional to $I$. Thus, for the flux $\phi$ linked with the circuit itself, we may write

$$\phi = LI \tag{4.104}$$

where $L$ is a constant known as the *self-inductance* of the circuit. We may thus say that the self-inductance of a circuit is the flux linkage with itself caused by the flow of unit current. The unit of self-inductance, as for mutual inductance, is the *henry* (symbol H).

If the current in the circuit varies with time, an e.m.f. $E$ will be induced, where

$$E = -\frac{d\phi}{dt} = -L\frac{dI}{dt} \tag{4.105}$$

The negative sign indicates that the induced e.m.f. tends to oppose the change of current.

In our discussion of mutual inductance we saw that the term flux linkage is somewhat ambiguous when the circuit conductor has appreciable cross-section, since the flux linked has different values for different paths round the circuit. The same problem arises in the consideration of self-inductance, but in a more acute form, since we can no longer evade the difficulty by assuming the conductor to have negligible area of cross-section. Such an assumption would lead to infinite flux density just outside the conductor and the self-inductance of the circuit would also become infinite.

The procedure for finding the flux linkage when current $I$ flows in the circuit is therefore as follows. We divide the total flux which links any part of the current into bundles $I\mathrm{d}\phi_1, I\mathrm{d}\phi_2 \ldots$ and we calculate the fraction of the total current with which each bundle is linked. Thus, let $I\mathrm{d}\phi_1$ be linked with fraction $k_1$, $I\,\mathrm{d}\phi_2$ with fraction $k_2$ and so on. Then, for the total flux, we have

$$\text{flux linkage} = \mathrm{Lt}_{n\to\infty}\, I\Sigma k_n\,\mathrm{d}\phi_n$$

and

$$L = \mathrm{Lt}_{n\to\infty}\, \Sigma k_n\,\mathrm{d}\phi_n \tag{4.106}$$

An alternative procedure which leads to the same result is to divide $I$ into fractions $p_1 I, p_2 I, \ldots$ and to calculate the fluxes $\phi_1, \phi_2, \ldots$ with which each filament is linked. Then

$$L = \mathrm{Lt}_{n\to\infty}\, \Sigma p_n \phi_n \tag{4.107}$$

We see from the above that the inductance of any circuit can be divided into two parts: a component $L_\mathrm{e}$ which arises from flux linked with the whole current and a component $L_\mathrm{i}$ resulting from flux which passes through the conductor and is therefore linked with only part of the current. Thus

$$L = L_\mathrm{e} + L_\mathrm{i} \tag{4.108}$$

$L_\mathrm{e}$ is known as the *external inductance* and $L_\mathrm{i}$ as the *internal inductance*. As a rule $L_\mathrm{e}$ is much greater than $L_\mathrm{i}$ but this is not always the case.

The above procedure does not completely overcome our difficulty since, if the cross-section of the conductor is appreciable, the e.m.fs induced in different paths round the circuit will not all be the same (§4.4.6) and eddy currents will flow in the conductor. These eddy currents will be added to the main current, with the result that the distribution of the total current over the cross-section of the conductor will be changed. This, in turn, will alter the magnetic field and hence the self-inductance of the circuit. The effects of eddy currents are most noticeable when the circuit is carrying high-frequency alternating current. It is then found that the eddy currents oppose the main current in the interior of the conductor and reinforce it near the surface. The overall result is that the total current appears to travel almost entirely through a thin layer near the surface of the conductor and the internal inductance then becomes very small (cf. §10.3.3).

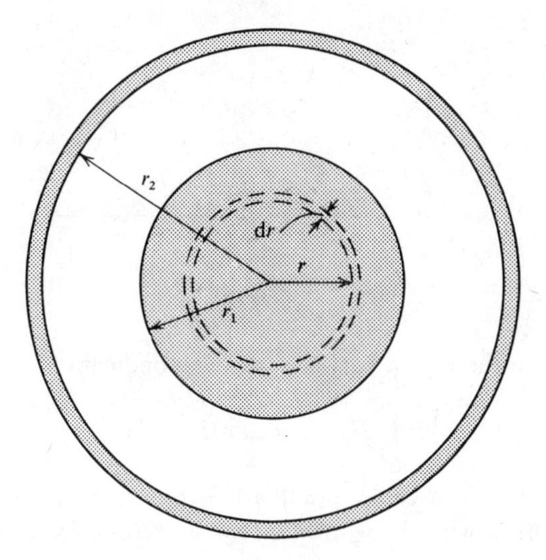


Fig. 4.18

### 4.5.4 The self-inductance of a coaxial cable

We consider the unit length of an infinitely long cable, whose cross-section is shown in fig. 4.18. The radius of the inner conductor is $r_1$, and $r_2$ is the internal radius of the outer conductor. The magnetic properties of the conductors will be assumed to be the same as those of free space (§7.2.1). It will also be assumed that the outer conductor has negligible thickness and that the current is uniformly distributed over the cross-sections of the conductors.

Consider first the region inside the inner conductor and take a circular path of radius $r$, centred on the axis. For total current $I$, the current linked by this path is $Ir^2/r_1^2$ and

$$\oint \boldsymbol{H}_r \cdot \mathrm{d}\boldsymbol{s} = 2\pi r H_r = Ir^2/r_1^2$$

or

$$H_r = Ir/2\pi r_1^2 \tag{4.109}$$

Remembering that the flux lines are circles round the axis, the flux between $r$ and $r+\mathrm{d}r$, for unit length of cable, is

$$\mu_0 H_r \mathrm{d}r = I\mu_0 r\,\mathrm{d}r/2\pi r_1^2 \tag{4.110}$$

This flux is linked with a fraction $r^2/r_1^2$ of the total current so, setting $I$ equal to unity, we have for the inner inductance $L_i$, by (4.106),

$$L_i = \frac{\mu_0}{2\pi r_1^4}\int_0^{r_1} r^3\,\mathrm{d}r = \frac{\mu_0}{8\pi} \tag{4.111}$$

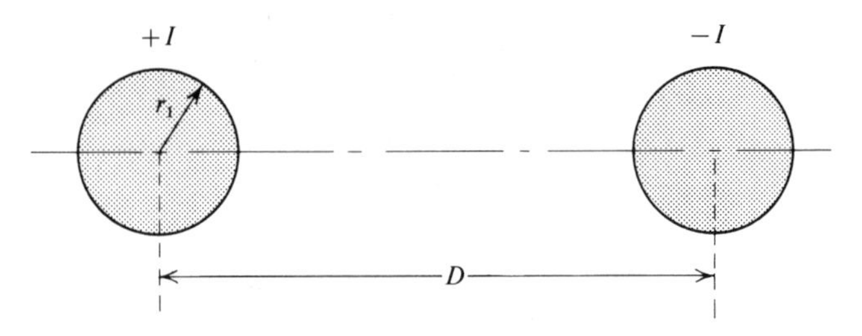


Fig. 4.19

Turning now to the region between the two conductors,

$$\oint_s H_r \cdot \mathrm{d}s = 2\pi r H_r = I \tag{4.112}$$

and, for unit length of cable, the flux between $r$ and $r+\mathrm{d}r$ is $\mu_0 I\,\mathrm{d}r/2\pi r$. This flux is linked with the whole current so, setting $I$ equal to unity, the external inductance is

$$L_e = \frac{\mu_0}{2\pi}\int_{r_2}^{r_1}\frac{\mathrm{d}r}{r} = \frac{\mu_0}{2\pi}\ln\frac{r_2}{r_1} \tag{4.113}$$

Finally, for the total inductance per unit length,

$$L = L_i + L_e = \frac{\mu_0}{2\pi}\left(\tfrac{1}{4} + \ln\frac{r_2}{r_1}\right)\ \mathrm{H\ m^{-1}} \tag{4.114}$$

### 4.5.5 The self-inductance of a parallel-wire circuit

We consider unit length of a circuit consisting of two long straight parallel circular conductors, each of radius $r_1$, separated by distance $D$ (fig. 4.19). Current $I$ flows through the two conductors in opposite directions and is assumed to be distributed uniformly over the cross-sections of the conductors. End effects are ignored.

The system is symmetrical, so the total inductance is twice that obtained by considering the current in one conductor only. Thus, for the left-hand conductor, the internal inductance per unit length will be the same as in the coaxial system (4.111) and

$$L_i' = \mu_0/8\pi \tag{4.115}$$

Outside the conductor, we have as before (4.112),

$$H_r = I/2\pi r \tag{4.116}$$

and, over the region from $r = r_1$ to $r = D - r_1$, the resulting magnetic flux links the whole of the current $I$. For radii greater than $D - r_1$ the fraction of the current linked decreases because part of the return current in the

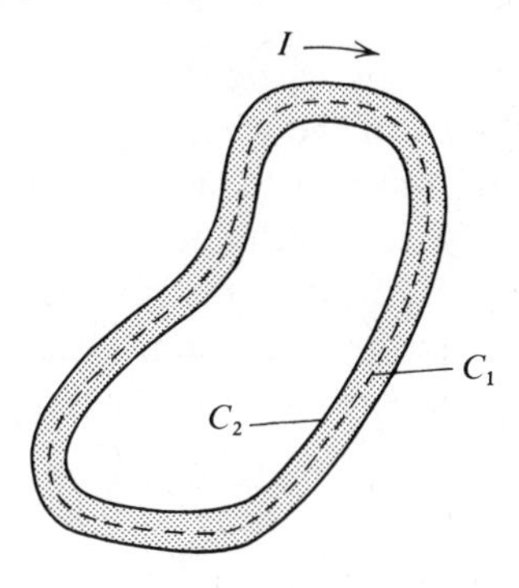


Fig. 4.20

right-hand conductor must be subtracted from the outgoing current. Finally, when $r$ is greater than $D+r_1$, the total current linked is zero. If $D$ is large compared with $r_1$, it will be a close approximation to the truth to take the fraction of current linked as unity from $r = r_1$ to $r = D$ and as zero for greater values of $r$. Then, putting $I$ equal to unity, the external inductance is

$$L'_e = \frac{\mu_0}{2\pi}\int_{r_1}^{D} \frac{dr}{r} = \frac{\mu_0}{2\pi} \ln \frac{D}{r_1} \tag{4.117}$$

So far, we have considered inductance resulting from current in the left-hand conductor only. Adding equal values for the right-hand conductor, the total inductance per unit length is

$$L = L_i + L_e = 2(L'_i + L'_e) = \frac{\mu_0}{\pi}\left(\tfrac{1}{4} + \ln \frac{D}{r_1}\right) \tag{4.118}$$

### 4.5.6 The self-inductance of an arbitrary loop of wire

In fig. 4.20 we suppose current $I$ to be flowing in a loop of wire of arbitrary shape. We assume the wire to have circular cross-section of radius $r_1$, which is very small compared with the radius of curvature of any part of the loop. An approximate formal expression for the self-inductance of the loop can be obtained as follows.

Under the conditions stated, the magnetic flux density at any point in the wire will be very nearly the same as that in a straight wire through which current $I$ is flowing. We have already seen (4.111) that the internal inductance per unit length of a straight wire is $\mu_0/8\pi$. Hence, for the loop, we can say that the internal inductance is very nearly equal to

$$L_i = \mu_0 l/8\pi \tag{4.119}$$

where $l$ is the total length of the loop.

Again, under the conditions stated, the flux density outside the wire will be very nearly the same as if the whole of the current had been concentrated along the mean contour of the wire $C_1$. The external inductance of the loop is then the flux resulting from unit concentrated current along $C_1$, which links the contour $C_2$ lying in the surface of the wire. However this flux linkage is simply the mutual inductance between the filamentary circuits $C_1$ and $C_2$. Hence we may write for the total self-inductance

$$L = \frac{\mu_0 l}{8\pi} + \begin{pmatrix}\text{mutual inductance between} \\ C_1 \text{ and } C_2\end{pmatrix} \tag{4.120}$$

## 4.6 Worked example

Two small flat coils, each 1 cm in diameter and having 50 turns of thin wire, are set in free space, with their centres 25 cm apart. The coils have their planes parallel, making an angle $\theta$ with the line joining their centres.

Find the mutual inductances between the two coils when $\theta$ has the value (*a*) 0, and (*b*) 90°. For what values of $\theta$ is the mutual inductance zero? (University of Sheffield, 1968.)

*Solution.* Let $F$ and $G$ be the two coils. We wish to calculate the flux linked with coil $G$ when unit current flows in coil $F$. Since the diameters of the coils are small compared with the distance between them, it is a reasonable approximation to replace $F$ by its equivalent dipole and to assume that the magnetic field over the whole of $G$ is the same as that at its centre.

The components of the field at $G$ are $H_r$ along the line joining the coils and $H_\theta$ at right angles to this line. Remembering that, in this example, $\theta$ is the angle between the line joining the coils and the planes of the coils, we have from (4.27) and (4.28),

$$\begin{aligned} H_r &= \frac{50 \times \pi(5 \times 10^{-3})^2}{2\pi(0.25)^3} \sin\theta \\ &= 0.04 \sin\theta \text{ A m}^{-1} \\ H_\theta &= \frac{50 \times \pi(5 \times 10^{-3})^2}{4\pi(0.25)^3} \cos\theta \\ &= 0.02 \cos\theta \text{ A m}^{-1} \end{aligned}$$

The flux densities resulting from these components are found by multiplying each by $\mu_0$.

When $\theta = 0$, $H_r$ is zero and the mutual inductance resulting from $H_\theta$ is

$$\begin{aligned} M_0 &= 0.02 \times 4\pi \times 10^{-7} \times 50 \times \pi(5 \times 10^{-3})^2 \\ &= 9.87 \times 10^{-11} \text{ H} \end{aligned}$$

When $\theta = 90°$, $H_\theta$ is zero and the mutual inductance resulting from $H_r$ is

$$M_{90} = 0.04 \times 4\pi \times 10^{-7} \times 50 \times \pi(5 \times 10^{-3})^2$$
$$= 1.97 \times 10^{-10}\ \text{H}$$

For an arbitrary value of $\theta$, $G$ presents an effective area $\pi(5 \times 10^{-3})^2 \cos\theta$ to $H_\theta$ and $\pi(5 \times 10^{-3})^2 \sin\theta$ to $H_r$. Moreover, the two components cause fluxes passing through $G$ in opposite directions. Hence, for zero mutual inductance,

$$0.02 \cos^2\theta - 0.04 \sin^2\theta = 0$$

giving

$$\theta = 35.3° \quad \text{or} \quad 144.7°$$

## 4.7 Problems

1. If

$$\boldsymbol{A} = A_x\boldsymbol{i} + A_y\boldsymbol{j} + A_z\boldsymbol{k} \quad \text{and} \quad \boldsymbol{B} = B_x\boldsymbol{i} + B_y\boldsymbol{j} + B_z\boldsymbol{k},$$

show that

$$\boldsymbol{A} \times \boldsymbol{B} = \begin{vmatrix} \boldsymbol{i} & \boldsymbol{j} & \boldsymbol{k} \\ A_x & A_y & A_z \\ B_x & B_y & B_z \end{vmatrix}$$

Hence find the value of the vector product

$$(2\boldsymbol{i} - 3\boldsymbol{j} - \boldsymbol{k}) \times (\boldsymbol{i} + 4\boldsymbol{j} - 2\boldsymbol{k})$$

2. Prove that, for any three vectors $\boldsymbol{A}$, $\boldsymbol{B}$ and $\boldsymbol{C}$,

$$\boldsymbol{A} \cdot (\boldsymbol{B} \times \boldsymbol{C}) = \begin{vmatrix} A_x & A_y & A_z \\ B_x & B_y & B_z \\ C_x & C_y & C_z \end{vmatrix}$$

Hence show that $\boldsymbol{A} \cdot (\boldsymbol{B} \times \boldsymbol{C}) = \boldsymbol{B} \cdot (\boldsymbol{C} \times \boldsymbol{A}) = \boldsymbol{C} \cdot (\boldsymbol{A} \times \boldsymbol{B})$

3. A plane square loop of wire, with length of side $l$, carries current $I$. Show that the magnetic field strength at the centre of the square has magnitude

$$H = 2(\sqrt{2})I/\pi l$$

4. The axis of a long cylindrical conductor, of radius $R_1$, lies along the $z$-axis of a rectangular coordinate system. The conductor contains a longitudinal cylindrical hole of radius $R_2$, whose axis is parallel to the $z$-axis and passes through the point $(0, b)$. Current $I$ flows along the conductor. Show that the magnetic field in the hole is uniform and find its value. (University of Cambridge, 1966.)

5. A long thin-walled non-magnetic conducting tube is situated in a uniform alternating magnetic field

$$H = H_0 \sin \omega t$$

with the axis of the tube parallel to the field. The tube has radius $R$, thickness $d$ and is made of material of resistivity $\rho$.

Show that, as a result of current induced in the tube, the magnitude of the field inside the tube is less than that outside by a factor

$$[1+(\omega\mu_0 Rd/2\rho)^2]^{\frac{1}{2}}$$

Flux in the material of the tube may be neglected.

6. Two parallel circuits of an overhead transmission line consist of four wires carried at the corners of a square. Find the flux, in webers per kilometre of circuit, linked with one of the circuits, when a current of 1 A flows in the other circuit.

The diameters of the wires may be assumed to be very small compared with the distance between them.

7. A toroid of rectangular cross-section has inner radius $R_1$, outer radius $R_2$ and axial width $b$. It is uniformly wound with a double layer of wire, with a total number of turns $N$. Show that its inductance is given by

$$L = (\mu_0 N^2 b/2\pi) \ln (R_2/R_1)$$

8. A solenoid of diameter $d$ and length $l$ is wound with an even number of layers containing a total of $N_1$ turns. A small circular coil of $N_2$ turns and area $A$ is mounted coaxially at the centre of the solenoid. It may be assumed that, when current flows in the solenoid, the resultant magnetic field is uniform over the area of the small coil. Show that the mutual inductance between the two coils is

$$M = \mu_0 N_1 N_2 A/(d^2+l^2)^{\frac{1}{2}}$$

9. A circular metal ring of resistivity $2.62\times 10^{-8}$ Ω m and density $2.7\times 10^3$ kg m$^{-3}$ is allowed to fall from rest, with its plane horizontal, through a radial magnetic field of flux density 0.1 tesla. At what velocity is the ring falling after 10 milliseconds, and what is its ultimate velocity?

Flux within the material of the ring may be neglected.

# 5

# Electric and magnetic fields in material media

## 5.1 Introduction

### 5.1.1 Statement of the problem

Hitherto our discussion has been limited to problems in which the whole of the field has been occupied by a single homogeneous isotropic medium and, in the cases of electrostatic and magnetic fields, this medium has been free space. We must now consider fields in which two or more media are present. We shall assume each medium to be homogeneous and isotropic and, in this chapter, we shall be concerned only with setting up the basic equations of the field and with very simple examples of their use.

The problem can be split into two parts, of which the first consists in setting up the equations relating to conditions in a single medium. So far as a conducting medium is concerned, this part of the problem has already been dealt with in chapter 2, since one medium differs from another only in having a different value of resistivity. In the cases of electrostatic and magnetic fields, however, we shall have to consider how fields in dielectric and magnetic materials differ from those in free space.

The second part of the problem involves the derivation of the boundary conditions that must be satisfied at the surface of separation between two media. These conditions, together with the equations relating to the interior of any one medium, provide sufficient physical information to enable one to solve any problem in principle. The resulting mathematical work may be very complex and in many cases numerical computation is necessary.

### 5.1.2 The basic experimental postulate

Our treatment of electrostatic fields in free space has been based on the assumption that such fields result from the presence of electric charges obeying the inverse-square law. Similarly, we have supposed magnetic fields in free space to be caused by electric currents in accordance with (4.6). When we come to fields in material substances, we must envisage the possibility that completely new agencies, which have no counterpart in free space, may cause electric and magnetic fields. For example, the existence of permanent magnetism might seem to support this view. If, in fact, such new agencies were discovered, we could hardly expect the equations established in chapters 3 and 4 to hold for the fields in material substances.

This is a matter which can only be settled by detailed comparison of theory with experiment, in investigations of the properties of a very wide range of substances. As a result of work of this kind it can be stated that there is no evidence to suggest the existence of any new agencies for the production of electric and magnetic fields, obeying laws which are fundamentally different from those which we have already considered. We know that atoms contain positively and negatively charged particles, obeying the inverse square law, and there is no reason to doubt that, in principle, the electric properties of materials can be explained in terms of these charges, coupled with the restrictions which the quantum theory places on their distribution and movements.

Similarly, magnetic behaviour can, in principle, be explained satisfactorily by assuming that minute circulating currents flow in atomic particles. We have already seen (§4.1.7) that a small current loop of this kind is equivalent to a magnetic dipole, so far as external magnetic effects are concerned. Thus it is quite immaterial whether we imagine the particles to be current loops or dipoles: in fact, we do not know their exact nature and, in the case of an electron for example, we refer to the property simply as *spin*. Experimental atomic physics provides abundant evidence for the existence of electron spin. For convenience we shall regard these atomic 'magnets' as small circulating currents whose magnetic effects are to be calculated by the methods of chapter 4. They are often referred to as amperian currents, since it was Ampère who first suggested that they might explain the magnetic behaviour of materials.

Summing up, we may say that electric and magnetic fields in materials are to be dealt with on the basis of the laws and relations discussed in chapters 3 and 4. We shall devote the present chapter to seeing how these laws and relations can be applied. For the time being we shall ignore effects resulting from permanent magnetism or the permanent polarization of insulators. These effects will be considered later (§7.2.10. §7.3.3).

## 5.2 Microscopic and macroscopic theories

### 5.2.1 Point values and average values

Our treatment of fields in free space was based on the vectors $\boldsymbol{E}$, $\boldsymbol{D}$, $\boldsymbol{B}$ and $\boldsymbol{H}$ and we were able to give a precise definition of the meaning of each of these quantities at any particular point. When we attempt to define these vectors at points in a material, we encounter a difficulty which we shall illustrate by reference to electric fields, though an exactly similar problem arises with magnetic fields.

We have said that electric fields in materials are caused by atomic charges in the material and by any external charges that may be present. Let us therefore consider what contribution to expect from the atomic charges. The distance between the centres of adjacent atoms in a solid is of the order of $10^{-10}$ m, so it is instructive to calculate the value of $E$ at a point midway between an electron (with negative charge of $1.6 \times 10^{-19}$ C) and an equal positive charge, when the two charges are separated by the above distance. It then appears that the order of magnitude of $E$ will be $10^{12}$ V m: that is to say it is vastly greater than any field strength likely to be caused by charges in free space outside the material. Our knowledge of atomic physics leads us to believe that a material contains very large numbers of these charges separated by atomic distances. Thus, we must conclude that, in any path through the material, we should expect to encounter fields of the order of $10^{12}$ V m, which are continually changing in magnitude and direction as we pass from point to point of the path. Moreover, since the atomic charges are in continual motion, there will be an equally violent variation with time of the field strength at a particular point. Since these very large field strengths are not observed on a macroscopic scale, we must conclude that the atomic fields largely cancel each other over any path comprising a considerable number of atoms though, as we shall see, the cancellation is not always complete.

With the above picture in mind, it will be clear that no useful meaning can be attached to the values of $\boldsymbol{E}$ and $\boldsymbol{D}$ at a particular point in a material at a particular instant of time. We must content ourselves with mean values which, in some fashion, are averaged over space and time. Exactly similar arguments lead us to the same conclusion with regard to the magnetic vectors $\boldsymbol{B}$ and $\boldsymbol{H}$.

### 5.2.2 Determination of the average values

A great deal is known about the properties of atoms, molecules and crystals and, ideally, one would wish to use this knowledge to calculate the average values of $\boldsymbol{E}$, $\boldsymbol{D}$, $\boldsymbol{B}$ and $\boldsymbol{H}$ in a material. Some progress in this direction has been made; it is possible, for example, to calculate the electrical properties of the alkali halides, in the form of single crystals, with fair accuracy. However, few materials that are of interest in electrical engineering are used in single-crystal form; many are composed of highly complex organic molecules, while still others are mixtures. It must be accepted that, for the foreseeable future, the electric and magnetic properties of most materials will have to be determined by experiment.

We have said that the vectors in which we are interested must be averaged with respect to space and time and we shall have to consider how this averaging process is to be carried out. Time-averaging presents no great difficulty, but we shall return to this matter later (§5.7.1). Space-averaging, however, can be carried out in more than one way. We might, for example, find the average value of $\boldsymbol{E}$ along a short length of a particular path in the material, or we might take the average over a small area at right angles to this path. In general, the results will be different in the two cases, so our definition of what we mean by $\boldsymbol{E}$ in the material depends on the choice that we make. To guide us we have two factors to consider. The definitions of $\boldsymbol{E}$ and $\boldsymbol{D}$ must be formulated in such a way that experiments to determine the values of these quantities can be carried out. Apart from this, the definitions should lead to equations which can be used to solve practical problems. In particular, it would be very convenient if relations similar to those developed for free space could be used when materials are present in the field.

### 5.2.3 Average values in conducting media

The equations governing the flow of current in a conducting medium have already been developed in chapter 2, but it was then assumed that the medium was completely homogeneous and no mention was made of its atomic structure. It is instructive to re-examine this matter in the light of the discussion set out in the preceding section, since this will afford guidance as to the way in which we should deal with electric and magnetic fields in materials.

Current in a conducting medium results from the motion of charged particles of atomic dimensions and, from what has already been said, the electric field in the space between these particles must vary erratically in magnitude and direction, between values which are vastly in excess of any field that could be applied to the medium as a whole. Let $\boldsymbol{E}_0$ be the field at any point in the medium and let us consider the potential difference between two points $A$ and $B$ whose distance apart is extremely large compared with atomic dimensions. For any path between $A$ and $B$ which does not pass through any of the charged particles, we have

$$V_A - V_B = \int_A^B \boldsymbol{E}_0 \cdot \mathrm{d}\boldsymbol{s} \tag{5.1}$$

$V_A - V_B$ is a quantity which can be measured experimentally in suitable circumstances. For example, if current is flowing along a uniform metal rod, the variation of potential along the rod can be measured with a potentiometer. When experiments of this kind are carried out, we find that the potential varies smoothly and is not subject to the violent fluctuations which must characterize $\boldsymbol{E}_0$. We thus conclude that, for any element $\mathrm{d}L$ of the path between $A$ and $B$, which is large compared with atomic dimensions though small on a macroscopic scale, there is an average value $\boldsymbol{E}$ for the electric field strength, which is not subject to erratic fluctuation. Using this value, (5.1) can be re-written as

$$V_A - V_B = \int_A^B \boldsymbol{E} \cdot \mathrm{d}\boldsymbol{L} \tag{5.2}$$

It is this average value $\boldsymbol{E}$, which can be deduced from experimental measurements, that is used throughout chapter 2.

Exactly similar considerations apply to the definition of the current density $\boldsymbol{J}_0$ in a medium. On an atomic scale the current density $\boldsymbol{J}_0$ must vary in a highly erratic manner from point to point. If, however, we choose an element of area which is large on an atomic scale though small from a macroscopic point of view, the fluctuations average out to give a mean value $\boldsymbol{J}$. For such an element $\mathrm{d}S$ with unit normal $\boldsymbol{n}$, we may write for the current passing through it

$$\mathrm{d}I = \boldsymbol{J} \cdot \boldsymbol{n} \, \mathrm{d}S$$

and, for the total current $I$ through any area $S$

$$I = \int_S \boldsymbol{J} \cdot \boldsymbol{n} \, \mathrm{d}S \tag{5.3}$$

It is this mean value $\boldsymbol{J}$, defined as above, that is used throughout chapter 2. It is a value which can be deduced from experimental measurements.

The above discussion can be summarized as follows. The values of $\boldsymbol{E}$ and $\boldsymbol{J}$ used in chapter 2 are average values taken over regions of space which are sufficiently large compared with atomic dimensions for erratic variations to have been evened out. The averaging process is carried out in quite different ways for the two vectors; for $\boldsymbol{E}$ we take the average along an element of path, while for $\boldsymbol{J}$ the average is for an element of area. The justification for these choices is that they lead to quantities which can be derived from experimental measurements. This is a necessary condition since, in the relation between $\boldsymbol{J}$ and $\boldsymbol{E}$

$$\boldsymbol{J} = \sigma \boldsymbol{E} \tag{5.4}$$

the conductivity $\sigma$ can only be determined by experiment.

## 5.3 The electrostatic field in a material medium

### 5.3.1 The nature of the medium

Materials can be divided into two broad classes: conductors and insulators (or dielectrics). In a conductor, electric charge can move freely so that, under equilibrium conditions, no electrostatic field can exist in such a material: any initial potential difference will cause the flow of current until the field strength has been reduced to zero. The whole of the conductor will then be at the same potential.

In a dielectric, the charged particles which exist in molecules are bound to those molecules and cannot move from one molecule to another. It is a matter of common observation that, in general, a piece of dielectric material does not, of itself, produce an electrostatic field on a macroscopic scale in the free space surrounding it. We thus conclude that the total positive charge in the atoms and molecules is exactly equal to the total negative charge and that the fields to which these charges give rise cancel each other at all external points whose distance from the material is greater than a few atomic diameters. This equality of positive and negative atomic charge is in accordance with our knowledge of atomic physics: it is a fact that we shall have to take into account in setting up our equations for fields in the material.

When a piece of dielectric material is placed in an electrostatic field produced by external charges, the positive and negative atomic charges may be displaced from their normal positions and the external fields to which they give rise will not then necessarily cancel. We shall see that this hypothesis provides a satisfactory explanation of the observed electric properties of insulators.

There is the further possibility that a piece of dielectric material may acquire a nett charge (e.g. by friction or by bombardment with electrons), so that it produces an external field independently of any external charges. This effect may be of considerable practical importance, but we shall not consider it further. One rarely has knowledge of the distribution of the additional charge, so calculation of the field to which it gives rise is impossible.

The division of all materials into conductors, on the one hand, and dielectrics on the other, is not completely satisfactory since almost all dielectrics possess some very small conductivity; that is, they contain a small number of mobile charged particles. If such a dielectric is left for a sufficiently long time in an external electrostatic field, the mobile charge will re-distribute itself in such a way as to bring the whole specimen to a uniform potential; the material will behave as a conductor. In the case of a good insulator, the re-distribution may take days or even years to approach completion. Thus, when we classify a substance as a dielectric, we mean that it does not permit appreciable re-distribution of charge in the period of time with which are are concerned.

### 5.3.2 The vectors $D$ and $E$ in an insulator

Summarizing the foregoing discussion, we may say:

(*a*) A dielectric contains very large numbers of positive and negative atomic charges, which are bound to individual atoms or molecules. The total quantities of positive and negative charge are equal.

(*b*) These charges produce local electric fields $\boldsymbol{E}_0$ and local electric displacements $\boldsymbol{D}_0$ which, at a particular point and at a particular instant of time, are related as they would be in free space by the equation

$$\boldsymbol{D}_0 = \epsilon_0 \boldsymbol{E}_0 \tag{5.5}$$

(*c*) Both $\boldsymbol{D}_0$ and $\boldsymbol{E}_0$ fluctuate wildly from point to point of the material and with time at any one point. Over distances which are large compared with atomic dimensions and over times large in comparison with the periods of movement of atomic particles, the fluctuations average out and it should therefore be possible to define mean values of $\boldsymbol{D}$ and $\boldsymbol{E}$.

(*d*) When a dielectric contains no excess charge on itself and is not under the influence of any electrostatic field produced by external charges, the average values of both $\boldsymbol{D}$ and $\boldsymbol{E}$ are zero (but see §7.3.3). When, however, an external field acts on the material, the atomic charges may be displaced from their normal positions in such a way that the average values of $\boldsymbol{D}$ and $\boldsymbol{E}$ are no longer zero. We must now consider how these average values can best be defined.

We have no reason to doubt that the equations established in chapter 3 for electrostatic fields in free space, hold also for the local fields in a material characterized by $\boldsymbol{D}_0$ and $\boldsymbol{E}_0$. Thus, if we consider any path in the material between two points $A$ and $B$, whose distance apart is very large in comparison with atomic dimensions, and if we suppose this path to be traversed in an instant of time during which $\boldsymbol{E}_0$ does not change, we may write for the potential difference between $A$ and $B$

$$V_A - V_B = \int_A^B \boldsymbol{E}_0 \cdot \mathrm{d}\boldsymbol{s} \tag{5.6}$$

We ought, perhaps, to specify that the path shall not pass through any of the atomic charged particles but, since these particles occupy only a minute fraction of the volume of an atom, this is not a serious limitation. In the same way, for any closed path, we have

$$\oint \boldsymbol{E}_0 \cdot \mathrm{d}\boldsymbol{s} = 0 \tag{5.7}$$

Moreover, (5.6) and (5.7) will be equally true for paths which lie partly inside and partly outside the material.

It would clearly be convenient if we could define our mean value $\boldsymbol{E}$ in the material in such a way that these equations remain true when $\boldsymbol{E}$ is substituted for $\boldsymbol{E}_0$ and this can be done as follows. We choose a length of line $\mathrm{d}\boldsymbol{L}$ passing through any point $P$ in the material, such that $\mathrm{d}\boldsymbol{L}$ is very large compared with atomic dimensions, but is nevertheless very small on a macroscopic scale. We then take the average value of the component of $\boldsymbol{E}_0$ along this line over the whole length $\mathrm{d}\boldsymbol{L}$ and we *define* this average value to be the component of $\boldsymbol{E}$ at $P$. The magnitude of the component will depend on the direction in which $\mathrm{d}\boldsymbol{L}$ lies and it will have a maximum value in some particular direction which is taken to be the direction of $\boldsymbol{E}$ at $P$.

From the above definition it follows at once that

$$\boldsymbol{E} \cdot \mathrm{d}\boldsymbol{L} = \int_0^{\mathrm{d}L} \boldsymbol{E}_0 \cdot \mathrm{d}\boldsymbol{s} \tag{5.8}$$

where $\mathrm{d}\boldsymbol{s}$ is an element of the path $\mathrm{d}\boldsymbol{L}$. Hence, from (5.6) and (5.7), for any path between points $A$ and $B$, whether wholly or partly in the material,

$$V_A - V_B = \int_A^B \boldsymbol{E} \cdot \mathrm{d}\boldsymbol{L} \tag{5.9}$$

and for any closed path,

$$\int_L \boldsymbol{E} \cdot \mathrm{d}\boldsymbol{L} = 0 \tag{5.10}$$

Turning now to the vector $\boldsymbol{D}$ in the material, this will have a local value $\boldsymbol{D}_0$ at any point, where

$$\boldsymbol{D}_0 = \epsilon_0 \boldsymbol{E}_0 \tag{5.11}$$

To decide how to average $\boldsymbol{D}_0$, we note that its most useful property is expressed in the equation

$$\oint_S \boldsymbol{D}_0 \cdot \boldsymbol{n} \mathrm{d}S = Q \tag{5.12}$$

where $Q$ is the total charge contained in any closed surface $S$. We therefore *define* $\boldsymbol{D}$ in the material to be the average value of $\boldsymbol{D}_0$ taken over a small element of area at right angles to the average electric field $\boldsymbol{E}$. The area is assumed to be macroscopically small, though large enough to ensure cancellation of fluctuations on an atomic scale. With this definition it follows that

$$\oint_S \boldsymbol{D} \cdot \boldsymbol{n} \, dS = Q \tag{5.13}$$

In (5.13) $Q$ is the total charge within $S$ and, if $S$ passes through or contains dielectric material, $Q$ must include atomic charges. However, since we are concerned with a macroscopic theory and since the electric properties of a material must be determined by experiments carried out on pieces of finite size, it is reasonable to stipulate that the surface $S$ must contain only complete molecules. But, as we have seen, the total charge on a molecule is zero. Hence, in (5.13), $Q$ is the total charge *outside* the material.

In the above discussion we have excluded the possibility that excess charge may reside within, or on the surface of an insulating material, though this can certainly happen in practice. There is no difficulty about including the effects of such charges if their positions and magnitudes are known, but we shall not pursue this matter.

### 5.3.3 The relation between $D$ and $E$ in a dielectric

Since $\boldsymbol{D}_0$ and $\boldsymbol{E}_0$ are related by (5.11), it might be thought that the same relation should hold for $\boldsymbol{D}$ and $\boldsymbol{E}$, but this is not the case. To understand why this should be so, we consider an idealized situation in which a piece of dielectric contains molecules which possess a permanent dipole moment (§3.6.6). It is assumed that the molecules are free to rotate but that, in the absence of an external field, their thermal agitation causes the axes of the dipoles to be orientated at random, as in fig. 5.1(*a*).

If now the material is placed in an electric field $\boldsymbol{E}$, the molecules will tend to turn so that their dipole axes lie in the direction of $\boldsymbol{E}$. For simplicity we have shown them completely aligned in fig. 5.1(*b*). It is now clear that, whatever may have been the case originally, the material is no longer electrically isotropic. Moreover, in defining the average values $\boldsymbol{D}$ and $\boldsymbol{E}$ in the material, we have averaged $\boldsymbol{E}$ along the line of the field and $\boldsymbol{D}$ over a surface at right angles to the field. It is hardly surprising that the presence of the dipole should have affected $\boldsymbol{D}$ and $\boldsymbol{E}$ quite differently.

If we consider the narrow rectangular path $ABCD$ in fig. 5.1(*b*), the line integral of $\boldsymbol{E}$ round this path must be zero. Hence, the average value of $\boldsymbol{E}$ along $AD$, outside the material, must be the same as that along $BC$, inside the material; the dipoles have made little difference. On the other hand, remembering that unit flux of $\boldsymbol{D}$ begins on each unit positive charge and ends on each unit negative charge, it is clear that the aligned dipoles must have added very greatly to the total flux of $\boldsymbol{D}$ within the material. Thus we

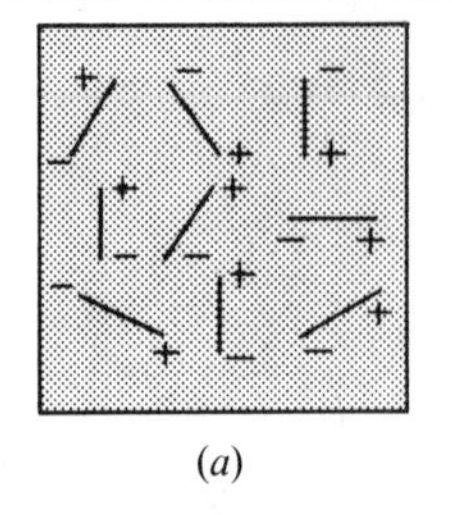

(a)

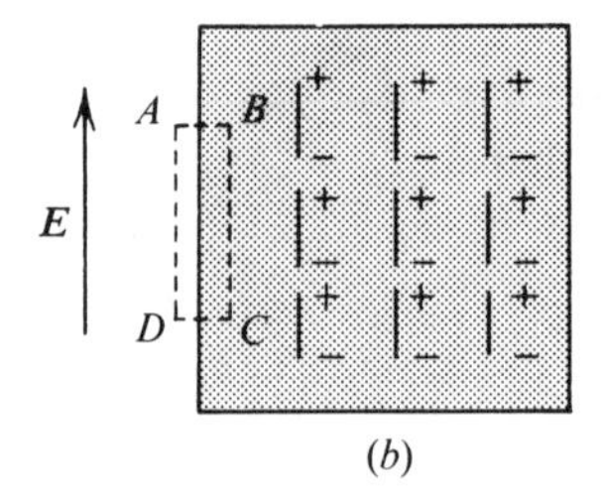


(b)

Fig. 5.1

conclude that the ratio of $\boldsymbol{D}$ to $\boldsymbol{E}$ in the material will be considerably greater than the corresponding ratio in free space $\epsilon_0$. We express this fact by writing

$$\boldsymbol{D} = \epsilon \boldsymbol{E} = \epsilon_r \epsilon_0 \boldsymbol{E} \tag{5.14}$$

where $\epsilon$ is the *permittivity* of the material. $\epsilon_r$ is termed the *relative permittivity* of the material. It is a pure number and has, in the past, been known as the *dielectric constant* or the *specific inductive capacity*. In general, $\epsilon_r$ cannot be calculated with any accuracy from the known atomic and molecular properties of a material: it must be found by applying a known uniform electric field $\boldsymbol{E}$ to a suitable specimen and measuring experimentally the resulting value of $\boldsymbol{D}$. The way in which such measurements can be carried out will be explained later (§7.3.2). Typical values of relative permittivity are given in table 5.1.

Table 5.1. *The relative permittivity of some common materials*

| Material | $\epsilon_r$ |
|---|---|
| Air (at atmospheric pressure) | 1.000 59 |
| Mica | 4.0–7.0 |
| Porcelain | 5.0–6.0 |
| Paraffin wax | 2.1–2.5 |
| Crown glass | 5.0–7.0 |
| Polystyrene | 2.4–2.8 |
| Fused quartz | 3.7–4.1 |
| Transformer oil | 2.24 |

### 5.3.4 The polarization vector $P$ and the susceptibility $\chi_e$

We have seen that the total displacement $\boldsymbol{D}$ in an insulating material is the sum of two components: one which would exist in the absence of any atomic or molecular charges and a second which can be ascribed directly to these charges. Thus, we may write formally,

$$\boldsymbol{D} = \epsilon_0 \boldsymbol{E} + \boldsymbol{P} \tag{5.15}$$

where $\boldsymbol{P}$ is known as the *electric polarization.* It represents the contribution to $\boldsymbol{D}$ made by atomic and molecular charges and can readily be shown to be the dipole moment per unit volume of the material.

Similarly it is sometimes convenient to write

$$\boldsymbol{P} = \epsilon_0 \chi_e \boldsymbol{E} \tag{5.16}$$

where $\chi_e$ is known as the *electric susceptibility* of the material. From (5.15),

$$\boldsymbol{D} = \epsilon_0(1+\chi_e)\boldsymbol{E} \tag{5.17}$$

and, from (5.14)

$$\epsilon = \epsilon_0 \epsilon_r = \epsilon_0(1+\chi_e) \tag{5.18}$$

or

$$\epsilon_r = 1+\chi_e \tag{5.19}$$

The reader should be aware of the definitions of $\boldsymbol{P}$ and $\chi_e$, since he may meet these quantities in the literature. We shall, however, make no further use of them in this book.

## 5.4 The magnetic field in a material medium

### 5.4.1 Introduction

Our treatment of the magnetic field in materials will follow lines closely similar to those adopted in dealing with electrostatic fields. It is therefore unnecessary to repeat many of the arguments which were set out in detail in §5.3.

We assume that the magnetic properties of a material result from the presence of atomic or molecular magnetic dipoles which, for convenience, we shall ascribe to minute circulating amperian currents. As explained in §5.1.2 the exact nature of the dipoles does not affect the macroscopic theory.

At any point in the material there will be a local magnetic field strength $\boldsymbol{H}_0$ and a local value of the magnetic flux density $\boldsymbol{B}_0$ produced partly by the dipoles and partly by any external field that may be present. As in free space,

$$\boldsymbol{B}_0 = \mu_0 \boldsymbol{H}_0 \tag{5.20}$$

Both $\boldsymbol{B}_0$ and $\boldsymbol{H}_0$ will vary, between very wide limits, with position and with time, so average values must be adopted in any useful macroscopic theory.

### 5.4.2 Definitions of $B$ and $H$ in a material

To define the value of $\boldsymbol{H}$ at any point $P$ in a material we proceed as we did in defining $\boldsymbol{E}$. We choose a line of length $\mathrm{d}\boldsymbol{L}$ passing through $P$, such that $\mathrm{d}\boldsymbol{L}$ is very large on an atomic scale but very small from a macroscopic point of view. We then take the average value of the component of $\boldsymbol{H}_0$ along this line over the whole length d$\boldsymbol{L}$ and we *define* this average value to be the component of $\boldsymbol{H}$ at $P$, in the direction d$\boldsymbol{L}$. The average value will depend on this direction and it will have a maximum in some particular direction which we take to be the direction of $\boldsymbol{H}$ at $P$. With this definition it follows that

$$\boldsymbol{H}\cdot \mathrm{d}\boldsymbol{L} = \int_0^{\mathrm{d}L} \boldsymbol{H}_0 \cdot \mathrm{d}\boldsymbol{s} \tag{5.21}$$

and, for any closed path, whether wholly or partially in the medium,

$$\oint_L \boldsymbol{H}\cdot \mathrm{d}\boldsymbol{L} = I \tag{5.22}$$

where $I$ is the total current linked with the path. Since we are establishing a macroscopic theory, we make the further stipulation that the path must not thread any of the amperian current loops in the molecules themselves. With this stipulation, the current $I$ in (5.22) is the current outside the medium plus any *macroscopic* current which may be flowing through the medium itself.

To define the vector $\boldsymbol{B}$ at the point $P$ we take an element of area d$S$ at right angles to $\boldsymbol{H}$ at the point. d$S$, though small on a macroscopic scale, is large compared with atomic dimensions. The average value of $\boldsymbol{B}$ at $P$ is then *defined* to be the average value of $\boldsymbol{B}_0$ over d$S$. It then follows that, for any arbitrary surface, $S$, whether wholly or partially in the medium,

$$\oint_S \boldsymbol{B}\cdot \boldsymbol{n}\,\mathrm{d}S = 0 \tag{5.23}$$

### 5.4.3 The relation between $B$ and $H$ in a material medium. Relative permeability

Because of the different ways in which $\boldsymbol{B}$ and $\boldsymbol{H}$ have been averaged in the medium, we cannot expect the relation between these quantities to be the same as that between $\boldsymbol{B}_0$ and $\boldsymbol{H}_0$. The physical explanation of this fact is similar to that given in connection with $\boldsymbol{D}$ and $\boldsymbol{E}$, and need not be repeated. To take account of the change, we write

$$\boldsymbol{B} = \mu \boldsymbol{H} = \mu_0 \mu_r \boldsymbol{H} \tag{5.24}$$

where $\mu$ is known as the *permeability* of the medium and has the same dimensions as $\mu_0$. $\mu_r$ is known as the *relative permeability* and is a pure number. It is a constant of the material which must be found by experiment.

### 5.4.4 Magnetization and magnetic susceptibility

The reader should be aware of certain other quantities which are sometimes used in the theory of magnetism.

The *magnetization vector* $\boldsymbol{M}$ is defined by the relation

$$\boldsymbol{M} = \frac{\boldsymbol{B}}{\mu_0} - \boldsymbol{H} \tag{5.25}$$

It corresponds to a splitting of the total density $\boldsymbol{B}$ into a component $\mu_0\boldsymbol{H}$ resulting from the externally applied field and a component $\mu_0\boldsymbol{M}$ caused by the internal magnetic dipoles.

The *magnetic susceptibility* $\chi_m$ is defined by the relation

$$\chi_m = \mu_r - 1 \tag{5.26}$$

The quantity $(\boldsymbol{B} - \mu_0\boldsymbol{H})$ is sometimes known as the *magnetic polarization* and is denoted by the symbol $\boldsymbol{J}$.

We shall make no further use of these quantities in this book.

## 5.5 Boundary conditions

### 5.5.1 Equations at a boundary

We have seen that the flow of current in a conducting medium can be specified in terms of the electric field $\boldsymbol{E}$ and the current density $\boldsymbol{J}$ and that these vectors are related by the equation

$$\boldsymbol{J} = \sigma\boldsymbol{E} \tag{5.27}$$

where $\sigma$ is the conductivity. Similarly, the electrostatic field in a dielectric involves the vectors $\boldsymbol{D}$ and $\boldsymbol{E}$, while the magnetic field in a material medium is expressed with the aid of $\boldsymbol{B}$ and $\boldsymbol{H}$. All these quantities have now been defined and it remains to consider the relations which must be satisfied at the boundary between two different isotropic media. The procedure to be followed in obtaining these relations is common to the three types of field and it is convenient to deal with them together, with the aid of fig. 5.2.

Let $P$ be a point on the boundary between the two media, which we distinguish by the figures 1 and 2. In fig. 5.2(*a*), let the directions of the current densities $\boldsymbol{J}_1$ and $\boldsymbol{J}_2$ at $P$ make angles $\alpha_1$ and $\alpha_2$ respectively with the normal to the boundary. If we consider a small area $\mathrm{d}S$ of the boundary surrounding $P$, the current through $\mathrm{d}S$ must be the same on the two sides. Hence

$$J_1 \cos\alpha_1 \,\mathrm{d}S = J_2 \cos\alpha_2 \,\mathrm{d}S$$

or

$$J_1 \cos\alpha_1 = J_2 \cos\alpha_2 \tag{5.28}$$

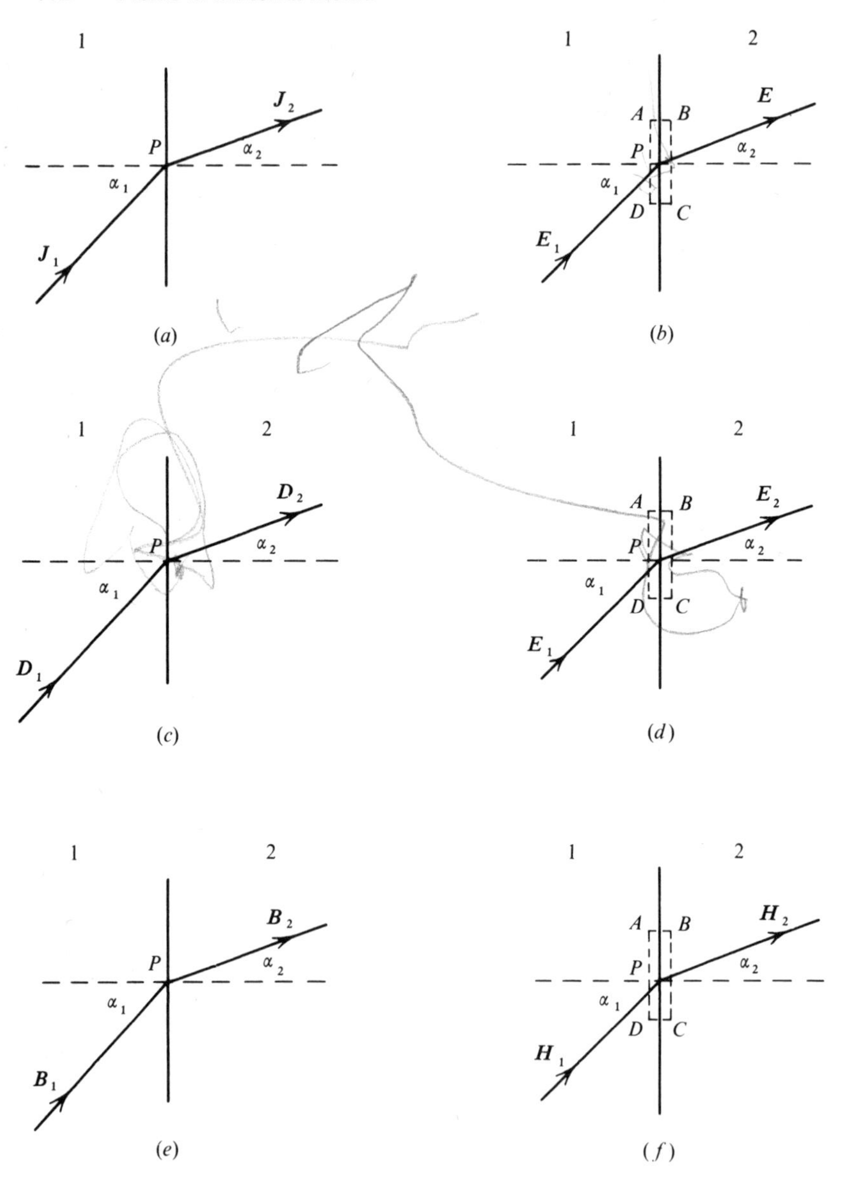


Fig. 5.2

Since we are dealing with isotropic media, $\boldsymbol{E}_1$ will have the same direction as $J_1$ and $\boldsymbol{E}_2$ as $\boldsymbol{J}_2$ so, in fig. 5.2(*b*), $\boldsymbol{E}_1$ and $\boldsymbol{E}_2$ make angles $\alpha_1$ and $\alpha_2$ respectively with the normal to the boundary. Consider a narrow rectangular path $ABCD$, in which the sides $AB$ and $CD$ are of negligible length. The line integral of $\boldsymbol{E}$ round this closed path must be zero, so

$$E_1 \sin \alpha_1 = E_2 \sin \alpha_2 \tag{5.29}$$

also
$$J_1 = \sigma_1 E_1 \quad \text{and} \quad J_2 = \sigma_2 E_2 \tag{5.30}$$

so, from (5.26), (5.27) and (5.28),

$$\frac{\tan \alpha_1}{\tan \alpha_2} = \frac{\sigma_1}{\sigma_2} = \frac{\rho_2}{\rho_1} \tag{5.31}$$

where $\rho_1$ and $\rho_2$ are the resistivities of the two media.

Turning now to the electrostatic field at the boundary between two dielectrics we have the situation depicted in figs. 5.2(*c*) and (*d*). In fig. 5.2(*c*) consider a small cylindrical gaussian surface surrounding $P$, with equal faces of area $\mathrm{d}S$ parallel to and very close to the boundary, connected by a curved surface of negligible area. Assuming no charge to reside on the boundary, the total flux of $\boldsymbol{D}$ into this surface from the left must be equal to the total outward flux on the right. Thus

$$D_1 \cos \alpha_1 \, \mathrm{d}S = D_2 \cos \alpha_2 \, \mathrm{d}S$$

or
$$D_1 \cos \alpha_1 = D_2 \cos \alpha_2 \tag{5.32}$$

In fig. 5.2(*d*), by the same argument as that used for fig. 5.2(*b*),

$$E_1 \sin \alpha_1 = E_2 \sin \alpha_2 \tag{5.33}$$

Also
$$D_1 = \epsilon_0 \epsilon_{r1} E_1 \quad \text{and} \quad D_2 = \epsilon_0 \epsilon_{r2} E_2 \tag{5.34}$$

Hence
$$\frac{\tan \alpha_1}{\tan \alpha_2} = \frac{\epsilon_{r1}}{\epsilon_{r2}} \tag{5.35}$$

Finally, in the case of a magnetic field, the flux of $\boldsymbol{B}$ passing through any small area of $\mathrm{d}S$ surrounding point $P$ in fig. 5.2(*e*) must be continuous as we go through the boundary, so

$$B_1 \cos \alpha_1 = B_2 \cos \alpha_2 \tag{5.36}$$

In fig. 5.2(*f*), the line integral of $\boldsymbol{H}$ round the closed path $ABCD$ must be zero, assuming no current to be flowing along the boundary, so

$$H_1 \sin \alpha_1 = H_2 \sin \alpha_2 \tag{5.37}$$

Also
$$B_1 = \mu_0 \mu_{r1} H_1 \quad \text{and} \quad B_2 = \mu_0 \mu_{r2} H_2 \tag{5.38}$$

Thus
$$\frac{\tan \alpha_1}{\tan \alpha_2} = \frac{\mu_{r1}}{\mu_{r2}} \tag{5.39}$$

### 5.5.2 Flux vectors

Our earlier discussion of fields in a single medium emphasized that $\boldsymbol{J}$, $\boldsymbol{E}$, $\boldsymbol{D}$, $\boldsymbol{B}$ and $\boldsymbol{H}$ were all flux vectors according to the definition of §2.2.2. Now that we are considering fields spanning more than one medium this statement is no longer true.

In the case of current flow, the essential property which makes $\boldsymbol{J}$ a flux vector (§2.4.2) is the fact that the quantity

$$I = \int_S \boldsymbol{J}\cdot\boldsymbol{n}\,\mathrm{d}S$$

depends only on the perimeter of the surface $S$ and is the same for all surfaces bounded by this perimeter. Applying this criterion to an element $\mathrm{d}S$ lying in the surface of separation between two media, $\boldsymbol{J}\cdot\boldsymbol{n}\,\mathrm{d}S$ must be the same on both sides of the element. From (5.28) we see that this is, in fact, the case, so $\boldsymbol{J}$ is a flux vector. However, from (5.27) it is clear that, if $\boldsymbol{J}$ is a flux vector, $\boldsymbol{E}$ can no longer be one for the two media. By exactly similar arguments we can see from (5.30) and (5.34) that $\boldsymbol{D}$ and $\boldsymbol{B}$ remain flux vectors, while $\boldsymbol{E}$ and $\boldsymbol{H}$ do not. At the surface of separation between two media, fluxes of $\boldsymbol{J}$, $\boldsymbol{D}$ and $\boldsymbol{B}$ are continuous while, unless the electrical and magnetic properties of the two media are identical, fluxes of $\boldsymbol{E}$ and $\boldsymbol{H}$ undergo discontinuous changes.

## 5.6 The physical meaning of the definitions of $J$, $E$, $D$, $B$ and $H$

The reader who has followed the foregoing description of the manner in which the various vectors relating to material media are to be defined may feel that, although tidy mathematical relations have been achieved, there has been a certain arbitrariness about the procedure. The following further explanation may assist him to gain a clearer understanding of the meaning of the equations that we have derived. For convenience we shall deal only with the case of the electrostatic field when dielectric media are present, but exactly similar considerations apply to current flow in conducting media and to magnetic fields in materials.

It is a matter of experimental observation that the large fluctuations in $\boldsymbol{D}$ and $\boldsymbol{E}$ which must exist on an atomic scale average out when we consider a region which is large compared with atomic dimensions. It is then possible to define average values for the magnitudes and direction of the two vectors in such a way that they can be related to charges outside the dielectric, without references to the atomic charges inside the dielectric, by (5.10), (5.13) and (5.14). However, (5.14) contains the unknown constant

$\epsilon_r$ which, for each material, can only be determined by experiment. Because of our inability to deal mathematically with the complex system of atomic charges which exists in a dielectric, we have arranged matters so that the effect of these charges is represented in our equations by the experimentally determined value of $\epsilon_r$.

We can look at this problem in a slightly different way. Using the above definitions of $\boldsymbol{D}$ and $\boldsymbol{E}$, we can suppose the total flux through any piece of dielectric, in a field caused by external charges, to be divided into 'tubes' of rectangular cross-section. These tubes will cut equipotential surfaces at right angles and, with a large number of such surfaces, the whole dielectric will be divided into blocks which are nearly rectangular, but have slightly curved surfaces. By increasing the number of tubes and surfaces, we can reduce the curvature to any desired extent. Assuming this to have been done, the electric field in each block is uniform, with a constant difference of potential between the faces which are normal to the field. Thus the conditions in the block are precisely those which, as we shall see later (§7.3.2), are always satisfied when we measure the permittivity of a material. What we have done is to divide the material into a large number of very small rectangular blocks and then to assert that each of these blocks behaves in the same way electrically, as the much larger rectangular block on which we have made experimental measurements under exactly the same conditions.

## 5.7 Practical problems

### 5.7.1 Soluble and insoluble problems

In the foregoing, emphasis has been placed on the fact that the definitions and equations which have been introduced are essentially macroscopic in nature. It follows that they can only be used to solve macroscopic problems. These include most of the problems which arise in engineering, since the engineer deals with pieces of material which are large on an atomic scale. By contrast, a physicist may wish to know something about the electric field acting on an individual atom or molecule when a voltage is applied between the opposite faces of a parallel-sided slab of material. Here, our theory will not directly help him. The theory is deliberately based on the experimentally measured properties of relatively large pieces of material, so it cannot, by itself, deal with problems on an atomic scale. On the other hand it is independent of any theory as to the structure of atoms and molecules.

The following example is instructive. We shall see later that the propagation of an electromagnetic wave in a dielectric is affected by the permit-

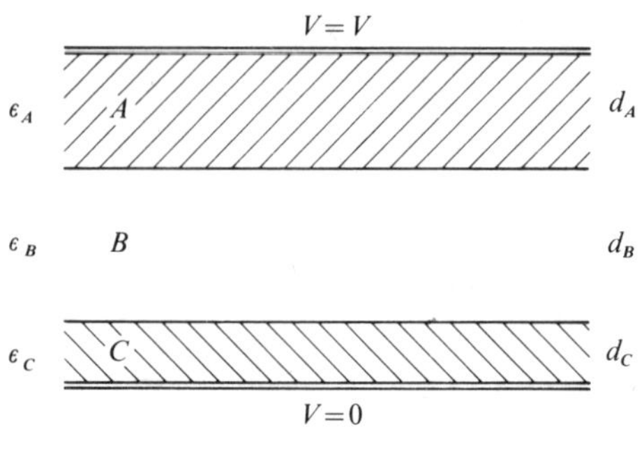


Fig. 5.3

tivity of the material. The question then arises whether the permittivity measured, for example, by an alternating-current bridge method is the relevant quantity to use in propagation problems.

The first point to be made here is that permittivity is unlikely to be the same at all frequencies. The relative permittivity of a material may result from various causes such as the displacement of electrons in an atom, the displacement of atoms in a molecule, or the rotation of whole molecules. The masses of the particles involved are different in the three cases and their reactions to alternating fields of different frequencies is also likely to be different. Hence, the permittivity must be measured at the frequency of the electromagnetic wave that we are considering. Provided this is done we should expect our macroscopic theory to be valid, so long as the wavelength of the waves is large compared with atomic dimensions. This is found to be the case for radio waves and for light. However, when we come to X-rays, the wavelength is of atomic dimensions and we should expect our theory to break down. As we know, new effects become manifest at these wavelengths.

### 5.7.2 Capacitors with more than one dielectric

In fig. 5.3 let two large parallel conducting plates be maintained at potentials $V = 0$ and $V = V$ respectively. The space between them is filled by three parallel slabs of dielectric of thicknesses $d_A$, $d_B$ and $d_C$, and relative permittivities $\epsilon_A$, $\epsilon_B$ and $\epsilon_C$ respectively. We wish to calculate the capacitance per unit area of the system, when edge effects are neglected.

Let $+Q$ and $-Q$ respectively be the charges per unit area on the two plates. By symmetry the displacement $\boldsymbol{D}$ between the plates must be normal to the plates and, since unit flux of $\boldsymbol{D}$ begins and ends on unit charge,

$$D = Q$$

and $D$ has the same value in all three dielectrics.

For dielectric $A$

$$E_A = D/\epsilon_0\epsilon_A = Q/\epsilon_0\epsilon_A$$

and the voltage across this dielectric is

$$V_A = E_A d_A = Qd_A/\epsilon_0\epsilon_A$$

For the whole system

$$V = V_A+V_B+V_C = \frac{Q}{\epsilon_0}\left[\frac{d_A}{\epsilon_A}+\frac{d_B}{\epsilon_B}+\frac{d_C}{\epsilon_C}\right]$$

and the capacitance $C$ per unit area is

$$C = \frac{Q}{V} = \frac{\epsilon_0}{\dfrac{d_A}{\epsilon_A}+\dfrac{d_B}{\epsilon_B}+\dfrac{d_C}{\epsilon_C}} \tag{5.40}$$

### 5.7.3 Dielectric strength

The *dielectric strength* of an insulator is defined to be the maximum value of the electric field that it can sustain without electrical breakdown. Typical values for a few materials are given in table 5.2.

Table 5.2. *Approximate dielectric strengths of some common materials*

| Material | Dielectric strength kV m$^{-1}$ |
|---|---|
| Air | 3000 |
| Mica | 100000 |
| Polystyrene | 20000 |
| Crown glass | 50000 |
| Porcelain | 10000 |
| Transformer oil | 12000 |

## 5.8 Worked example

In a coaxial cylindrical cable, the radius of the inner conductor is $a$ and the space between the inner and outer conductor is filled with a material whose safe working dielectric strength is $E_m$. Find the minimum internal radius of the outer conductor, if a voltage $V$ is to be applied between the two conductors.

*Solution.* Let $+Q$ and $-Q$ respectively be the charges per unit length of the conductors when voltage $V$ is applied. The total flux of $\boldsymbol{D}$ per unit length is equal to $Q$ and is at right angles to the conductors. Thus the magnitude of $\boldsymbol{D}$ at radius $r$ between the conductors is

$$D_r = Q/2\pi r \tag{5.41}$$

Hence $$E_r = Q/2\pi r\epsilon$$

where $\epsilon = \epsilon_0 \epsilon_r$ is the permittivity of the dielectric. The field strength will be greatest when $r = a$, and its value then can be equated to $E_m$. Thus

$$E_m = Q/2\pi a\epsilon \tag{5.42}$$

If $b$ is the radius of the outer conductor, we have from (5.41)

$$V = \int_a^b E_r \, dr = \frac{Q}{2\pi\epsilon} \ln \frac{b}{a} \tag{5.43}$$

Finally, from (5.42) and (5.43)

$$V/E_m = a \ln b/a$$

or $$b = a\, e^{V/aE_m} \tag{5.44}$$

## 5.9 Problems

1. A point $P$ in an electrostatic system is situated on a plane interface between air ($\epsilon_r = 1$) and glass ($\epsilon_r = 4$). There is no free charge on this interface.

In the vicinity of $P$, the electric displacement $\boldsymbol{D}$ in the air is inclined at an angle of 30° to the normal to the interface and the flux density of $\boldsymbol{D}$ is 0.25 micro-coulomb $m^{-2}$. What is the flux density of $\boldsymbol{D}$ in the glass and what is its inclination to the normal?

2. A voltage $V$ is applied between two large parallel plane conductors in air ($\epsilon_r = 1$). When the conductors are a distance $d$ apart, incipient dielectric breakdown of the air is found to occur and, in an attempt to cure this, half of the space between the conductors is filled with a plane glass sheet ($\epsilon_r = 4$), of thickness $d/2$. The dielectric strength of glass is very much higher than that of air and it may be assumed to have infinite resistivity.

Suggest the probable outcome of this change.

3. A condenser bushing incorporates two thin coaxial conducting tubes as shown in fig. 5.4. The spaces between the tubes, and between the inner tube and the central conductor, are filled with the same insulating material, which has uniform properties. The outer tube is earthed, the central conductor is maintained at a constant high potential, and the tube between them is insulated. It may be assumed that the electrostatic lines of force lie in planes perpendicular to the common axis of the conductors, and that there are no lines of force outside the insulating material.

Show that the maximum potential gradient in the system is reduced to the lowest possible value when $L$ and $D$ are given in centimetres by

$$L = 10D \quad \text{and} \quad \frac{2}{D} + \ln \frac{4}{D} = 1$$

(University of Cambridge, 1956.)

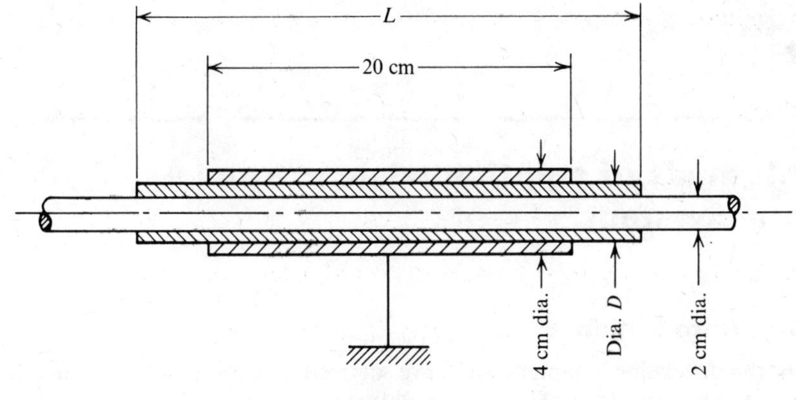


Fig. 5.4

4. In a parallel-plate capacitor the plates are distant $d$ apart and the space between them is filled with a dielectric whose relative permittivity varies linearly from $e_{r1}$ at one plate to $\epsilon_{r2}$ at the other ($\epsilon_{r2} > \epsilon_{r1}$). Show that the capacitance per unit area is

$$\frac{\epsilon_0(\epsilon_{r2}-\epsilon_{r1})}{d \ln(\epsilon_{r2}/\epsilon_{r1})}$$

# 6

# Methods of solution when σ, ε and μ are constant

## 6.1 Introduction

In the foregoing chapters we have derived equations which must be satisfied by all electric and magnetic fields, whether in free space or in space occupied by more than one medium. We have also obtained equations governing the flow of current in a conducting medium. We have assumed that each medium is homogeneous and isotropic and that the permittivity $\epsilon$, the permeability $\mu$ and the conductivity $\sigma$ are constants. These restrictions are assumed in the present chapter also. Such media may be said to be *linear* because $\boldsymbol{D}$ is proportional to $\boldsymbol{E}$, $\boldsymbol{B}$ to $\boldsymbol{H}$, and $\boldsymbol{J}$ to $\boldsymbol{E}$. *Non-linear* media will be considered in chapter 7.

Nearly all of the equations that we have derived have been in integral form, involving line integrals along particular paths or surface integrals over particular areas. Such equations can be used to solve problems only if we have additional information about the way in which the quantities under the integral sign vary along the path or over the surface. In practice this means that we can deal only with systems possessing planar, cylindrical or spherical symmetry. If we are to deal with more complicated systems we must transform the equations to differential forms which relate to conditions at a particular point in the system.

## 6.2 The equations of Poisson and Laplace

### 6.2.1 Electrostatic fields

We wish to find a differential form for the equation

$$\oint_S \boldsymbol{D}\cdot\boldsymbol{n}\,\mathrm{d}S = \Sigma Q \tag{6.1}$$

where $\Sigma Q$ is the total charge inside the closed surface $S$ (§3.4). However, $Q$ is not a convenient variable for our purpose, since the charge may be distributed throughout a considerable volume whereas we are seeking an

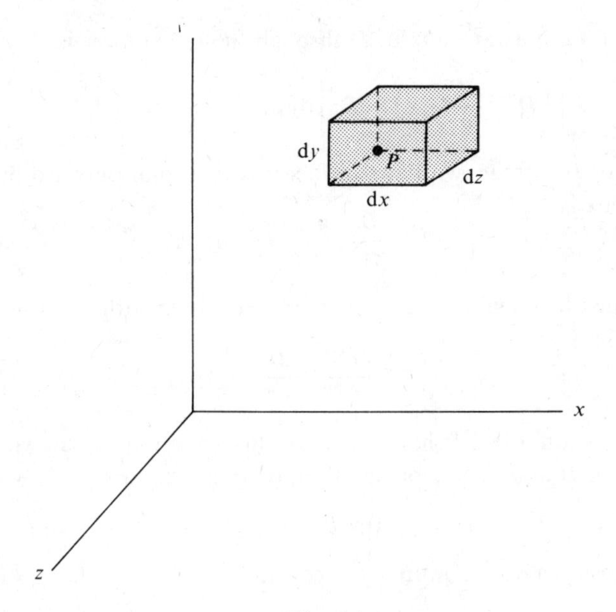


Fig. 6.1

equation connecting quantities at a particular point. We therefore write

$$\Sigma Q = \int_v \rho \, \mathrm{d}v$$

where $\rho$ is the *charge density* (i.e. the charge per unit volume) in the small element of volume $\mathrm{d}v$ and the integration is to be taken throughout the volume $v$ bounded by the closed surface $S$. Normally, $\rho$ will vary from point to point of this volume. Equation (6.1) now becomes

$$\oint_S \boldsymbol{D} \cdot \boldsymbol{n} \, \mathrm{d}S = \int_v \rho \, \mathrm{d}v \tag{6.2}$$

Turning now to fig. 6.1, we use rectangular coordinates and take as our element of volume a rectangular box with sides of length $\mathrm{d}x$, $\mathrm{d}y$ and $\mathrm{d}z$. The corner $P$ of the box has coordinates $(x, y, z)$.

The displacement $\boldsymbol{D}$ has components $D_x$, $D_y$ and $D_z$ and we first consider the two faces of the box perpendicular to the $x$-axis. The area of each of these faces is $\mathrm{d}y\mathrm{d}z$ and if $D_x$ is the component of $\boldsymbol{D}$ at the face passing through $P$, the value of this component at the opposite face will be

$$D_x + \frac{\partial D_x}{\partial x} \mathrm{d}x$$

Hence the total outward flux of $\boldsymbol{D}$ through these two faces is

$$\left[\left(D_x + \frac{\partial D_x}{\partial x}\mathrm{d}x\right) - D_x\right]\mathrm{d}y\,\mathrm{d}z = \frac{\partial D_x}{\partial x}\,\mathrm{d}x\,\mathrm{d}y\,\mathrm{d}z$$

Similarly for the other two pairs of faces; so the total outward flux of $\boldsymbol{D}$ from the box is

$$\left(\frac{\partial D_x}{\partial x} + \frac{\partial D_y}{\partial y} + \frac{\partial D_z}{\partial z}\right)\mathrm{d}x\,\mathrm{d}y\,\mathrm{d}z$$

But this must be equal to the total charge $\rho\,\mathrm{d}x\,\mathrm{d}y\,\mathrm{d}z$ within the box, so

$$\frac{\partial D_x}{\partial x} + \frac{\partial D_y}{\partial y} + \frac{\partial D_z}{\partial z} = \rho \tag{6.3}$$

The quantity on the left-hand side of this equation is known as the *divergence* of $\boldsymbol{D}$ and, in vector notation, (6.3) is written

$$\operatorname{div} \boldsymbol{D} = \rho \tag{6.4}$$

For a material whose permittivity is constant

$$\boldsymbol{D} = \epsilon_{\mathrm{r}}\epsilon_0 \boldsymbol{E} \tag{6.5}$$

and substitution in (6.4) gives

$$\operatorname{div} \boldsymbol{E} = \frac{\partial E_x}{\partial x} + \frac{\partial E_y}{\partial y} + \frac{\partial E_z}{\partial z} = \frac{\rho}{\epsilon_{\mathrm{r}}\epsilon_0} \tag{6.6}$$

We have already seen (§3.23) that $\boldsymbol{E}$ and the potential $V$ are connected by the relation

$$\boldsymbol{E} = -\operatorname{grad} V$$

or

$$E_x = -\frac{\partial V}{\partial x}, \quad E_y = -\frac{\partial V}{\partial y}, \quad E_z = -\frac{\partial V}{\partial z}$$

Substituting these values in (6.6) we get *Poisson's equation*

$$\frac{\partial^2 V}{\partial x^2} + \frac{\partial^2 V}{\partial y^2} + \frac{\partial^2 V}{\partial z^2} = -\frac{\rho}{\epsilon_{\mathrm{r}}\epsilon_0} \tag{6.7}$$

In the special case where $\rho$ is zero throughout the space considered, this reduces to

$$\frac{\partial^2 V}{\partial x^2} + \frac{\partial^2 V}{\partial y^2} + \frac{\partial^2 V}{\partial z^2} = 0 \tag{6.8}$$

This is *Laplace's equation*, which finds application in the theory of the flow of heat or of fluids and in the determination of stress in elastic solids, as well as in electromagnetic theory.

### 6.2.2 Magnetic fields

We have seen (§4.1.9) that the magnetic flux density $\boldsymbol{B}$ is a flux vector and that lines of $\boldsymbol{B}$ are always closed loops; isolated magnetic poles on which such lines might begin or end do not exist. It follows that the total outward flux of $\boldsymbol{B}$ through any closed surface must always be zero, or

$$\oint_S \boldsymbol{B}\cdot\boldsymbol{n}\,\mathrm{d}S = 0 \tag{6.9}$$

Repeating the arguments of the previous section we may re-write this relation as

$$\operatorname{div} \boldsymbol{B} = 0 \tag{6.10}$$

Furthermore, in regions where no current flows and where the permeability is constant, we have

$$\boldsymbol{B}/\mu_0\mu_r = \boldsymbol{H} = -\operatorname{grad} U \tag{6.11}$$

where $U$ is the magnetic scalar potential and

$$H_x = -\frac{\partial U}{\partial x}, \quad H_y = -\frac{\partial U}{\partial y}, \quad H_z = -\frac{\partial U}{\partial z}$$

Substituting in (6.10), we get

$$\frac{\partial^2 U}{\partial x^2} + \frac{\partial^2 U}{\partial y^2} + \frac{\partial^2 U}{\partial z^2} = 0 \tag{6.12}$$

which is Laplace's equation for a magnetic field.

### 6.2.3 Current flow in a conducting medium

We assume the currents to be steady, so that $\boldsymbol{J}$ does not vary with time, and we consider only that part of the medium where there are no sources and sinks. Then, for any closed surface,

$$\oint_S \boldsymbol{J}\cdot\boldsymbol{n}\,\mathrm{d}S = 0 \tag{6.13}$$

which may be written as

$$\operatorname{div} \boldsymbol{J} = 0 \tag{6.14}$$

If the conductivity $\sigma$ of the medium is constant,

$$\boldsymbol{J} = \sigma\boldsymbol{E}$$

and

$$\operatorname{div} \boldsymbol{E} = 0 \tag{6.15}$$

Also

$$\boldsymbol{E} = -\operatorname{grad} V$$

so

$$\frac{\partial^2 V}{\partial x^2} + \frac{\partial^2 V}{\partial y^2} + \frac{\partial^2 V}{\partial z^2} = 0 \tag{6.16}$$

as in the electrostatic field.

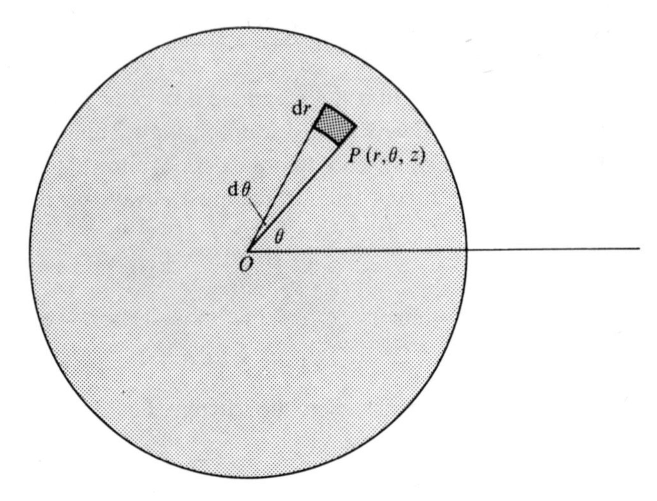


Fig. 6.2

### 6.2.4 Cylindrical and spherical coordinates

To obtain Laplace's equation in cylindrical coordinates $(r, \theta, z)$ we can make the substitutions

$$x = r\cos\theta, \quad y = r\sin\theta$$

and convert (6.16) by the ordinary techniques of partial differentiation. Alternatively, we can take an elementary volume of different form, as indicated in fig. 6.2, where the $z$-axis is taken to be normal to the plane of the diagram, but is not shown. The point $P$ has coordinates $(r, \theta, z)$.

The total outward flux of $\boldsymbol{D}$ through the two faces normal to $r$ is

$$\left(D_r + \frac{\partial D_r}{\partial r}\,\mathrm{d}r\right)(r+\mathrm{d}r)\,\mathrm{d}\theta\,\mathrm{d}z - D_r r\,\mathrm{d}\theta\,\mathrm{d}z = D_r\,\mathrm{d}r\,\mathrm{d}\theta\,\mathrm{d}z + \frac{\partial D_r}{\partial r}\,\mathrm{d}r\,\mathrm{d}\theta\,\mathrm{d}z$$

For the two faces normal to $r\,\mathrm{d}\theta$ we have

$$\left(D_\theta + \frac{1}{r}\frac{\partial D_\theta}{\partial \theta}\,r\,\mathrm{d}\theta\right)\mathrm{d}r\,\mathrm{d}z - D_\theta\,\mathrm{d}r\,\mathrm{d}z = \frac{\partial D_\theta}{\partial \theta}\,\mathrm{d}r\,\mathrm{d}\theta\,\mathrm{d}z$$

For the two faces normal to $z$ the outward flux is

$$\left(D_z + \frac{\partial D_z}{\partial z}\,\mathrm{d}z\right)r\,\mathrm{d}\theta\,\mathrm{d}r - D_z r\,\mathrm{d}\theta\,\mathrm{d}r = r\frac{\partial D_z}{\partial z}\,\mathrm{d}r\,\mathrm{d}\theta\,\mathrm{d}z$$

The charge within the element of volume is $\rho r\,\mathrm{d}r\,\mathrm{d}\theta\,\mathrm{d}z$ so

$$\operatorname{div}\boldsymbol{D} = \frac{1}{r}D_r + \frac{\partial D_r}{\partial r} + \frac{1}{r}\frac{\partial D_\theta}{\partial \theta} + \frac{\partial D_z}{\partial z} = \rho \qquad (6.17)$$

To obtain Laplace's equation we substitute

$$D = \epsilon_0 \epsilon_r E, \quad \rho = 0$$

$$E_r = -\frac{\partial V}{\partial r}, \quad E_\theta = -\frac{1}{r}\frac{\partial V}{\partial \theta}, \quad E_z = -\frac{\partial V}{\partial z}$$

to give

$$\frac{1}{r}\frac{\partial V}{\partial r} + \frac{\partial^2 V}{\partial r^2} + \frac{1}{r^2}\frac{\partial^2 V}{\partial \theta^2} + \frac{\partial^2 V}{\partial z^2} = 0 \tag{6.18}$$

By a similar procedure, Laplace's equation in spherical coordinates can be found, giving

$$\frac{\partial^2 V}{\partial r^2} + \frac{1}{r^2}\frac{\partial^2 V}{\partial \theta^2} + \frac{1}{r^2 \sin^2\theta}\frac{\partial^2 V}{\partial \phi^2} + \frac{2}{r}\frac{\partial V}{\partial r} + \frac{\cot\theta}{r^2}\frac{\partial V}{\partial \theta} = 0 \tag{6.19}$$

### 6.2.5 Properties of Laplace's equation

The following properties of Laplace's equation are of interest.

(*a*) If $V_1, V_2, \ldots, V_k$ are solutions of Laplace's equation, then

$$V = A_1 V_1 + A_2 V_2 + \ldots + A_k V_k$$

where the $A$s are arbitrary constants, is also a solution. This result follows at once when the expression for $V$ is substituted in the equation.

(*b*) In that part of a field to which Laplace's equation applies (i.e. $\rho = 0$), there can be neither a maximum nor a minimum of $V$.

This follows from the fact that at a maximum the partial derivatives of $V$ with respect to $x$, $y$ and $z$ must all be negative, while at a minimum they must all be positive. Since their sum is zero, there can be neither a maximum nor a minimum value of $V$. The potential must have its greatest and least values at points on the boundary, where there are charges.

A further deduction from this result is that, within a closed conducting surface which contains no charge, the electric field $\boldsymbol{E}$ must be zero at all points. The conducting surface will be an equipotential so, since maxima and minima of $V$ are excluded, $V$ must remain constant along any path from one point of the surface to another. But, if $V$ is constant along a path, $\boldsymbol{E}$ is zero (§3.6.4).

(*c*) A solution of Laplace's equation which also satisfies the boundary conditions is unique: for given boundary conditions there is only one distribution of $V$ that will satisfy both Laplace's equation and these conditions.

Analytical proofs of this statement are given in more advanced text books; here we shall rely on the following simple argument. If possible, let

$$V = \Phi_1 \quad \text{and} \quad V = \Phi_2$$

where $\Phi_1$ and $\Phi_2$ are functions of the coordinates, represent two different solutions of the equation. Then, from (*a*) above,

$$V = \Phi_1 - \Phi_2$$

is also a solution.

At the boundary of the field, $V$ has a prescribed value at all points, so $\Phi_1$ and $\Phi_2$ must have the same values at these points. Thus $(\Phi_1 - \Phi_2)$ must be zero at all points on the boundary. If $(\Phi_1 - \Phi_2)$ is not zero at all other points of the field, a maximum or minimum of $(\Phi_1 - \Phi_2)$ must exist. However, we have seen in (*b*) above that this is not possible. We therefore conclude that $\Phi_1$ and $\Phi_2$ are identical and that only one solution of Laplace's equation can satisfy the boundary conditions.

## 6.3 Methods of solving Laplace's equation

### 6.3.1 The general analytical method

Dealing first with the equation in rectangular coordinates

$$\frac{\partial^2 V}{\partial x^2} + \frac{\partial^2 V}{\partial y^2} + \frac{\partial^2 V}{\partial z^2} = 0$$

we assume a solution of the form

$$V(x, y, z) = X(x)\, Y(y)\, Z(z) \tag{6.20}$$

where each of the functions $X$, $Y$ and $Z$ is a function of one variable only. Substituting in (6.20) we get

$$\frac{\partial^2 X}{\partial x^2}\, YZ + \frac{\partial^2 Y}{\partial y^2}\, XZ + \frac{\partial^2 Z}{\partial z^2}\, XY = 0$$

or, diving by $XYZ$,

$$\frac{1}{X}\frac{\partial^2 X}{\partial x^2} + \frac{1}{Y}\frac{\partial^2 Y}{\partial y^2} + \frac{1}{Z}\frac{\partial^2 Z}{\partial z^2} = 0$$

The first term is a function of $x$ only, the second of $y$ only and the third of $z$ only. The sum of the three must be zero and the only way of satisfying this requirement for all values of $x$, $y$ and $z$ is for each of the three terms separately to be equal to a constant. We thus write

$$\frac{\partial^2 X}{\partial x^2} + a^2 X = 0 \tag{6.21}$$

$$\frac{\partial^2 Y}{\partial y^2} + b^2 Y = 0 \tag{6.22}$$

$$\frac{\partial^2 Z}{\partial z^2} + c^2 Z = 0 \tag{6.23}$$

where $a^2$, $b^2$ and $c^2$ are arbitrary constants except that

$$a^2+b^2+c^2 = 0 \tag{6.24}$$

$a^2$, $b^2$ and $c^2$ may be either positive or negative but, to satisfy (6.24), one of them must be of opposite sign to the other two. If $a^2$ is positive, the solution of (6.21) is either

$$X = a \sin x \quad \text{or} \quad X = a \cos x \tag{6.25}$$

while if $a^2$ is negative, we have

$$X = a \sinh x \quad \text{or} \quad X = a \cosh x \tag{6.26}$$

and similarly for $y$ and $z$.

Reverting now to (6.20), we see that any triple product of one of the solutions (6.25) or (6.26) and similar solutions for $y$ and $x$ is a solution of Laplace's equation. Moreover, by giving different values to two of the constants $a$, $b$ and $c$ (the third is fixed by (6.24)) we can obtain a doubly infinite set of such solutions and, as we have already seen, the sum of any number of these solutions will itself be a solution. The final problem is to build up a sum of solutions that will satisfy the given boundary conditions, in much the same way that one builds up a Fourier series to represent a curve of arbitrary shape. In the present instance the boundary conditions will normally be set by the given potentials of conductors of given shape.

The same general method of attack can be applied to problems expressed in cylindrical or spherical coordinates, where we should assume solutions of the form

$$V = R(r)\Theta(\theta)\,Z(z) \quad \text{or} \quad V = R(r)\Theta(\theta)\,\Phi(\phi)$$

In these cases the solutions will often involve Bessel or Legendre functions.

We shall not pursue this matter since, for the purpose of this book, it is more important that the reader should gain a simple picture of the method, than that he should become immersed in the mathematical details of a particular problem. The mathematical complexity is often considerable and, unless the boundary conditions are relatively simple, the problem may well be insoluble. To illustrate the method we choose a problem in which the algebra is straightforward.

Suppose an uncharged infinite cylinder of dielectric of permittivity $\epsilon$ and radius $r_0$ to be placed with its axis at right angles to a uniform electrostatic field of magnitude $E_0$, which already exists in free space, of permittivity $\epsilon_0$. We may suppose this uniform field to be caused by the application of a potential difference between two very large parallel plane conductors which are a great distance apart. The cylinder will disturb the field in its vicinity but we suppose the field to remain uniform and of magnitude $E_0$

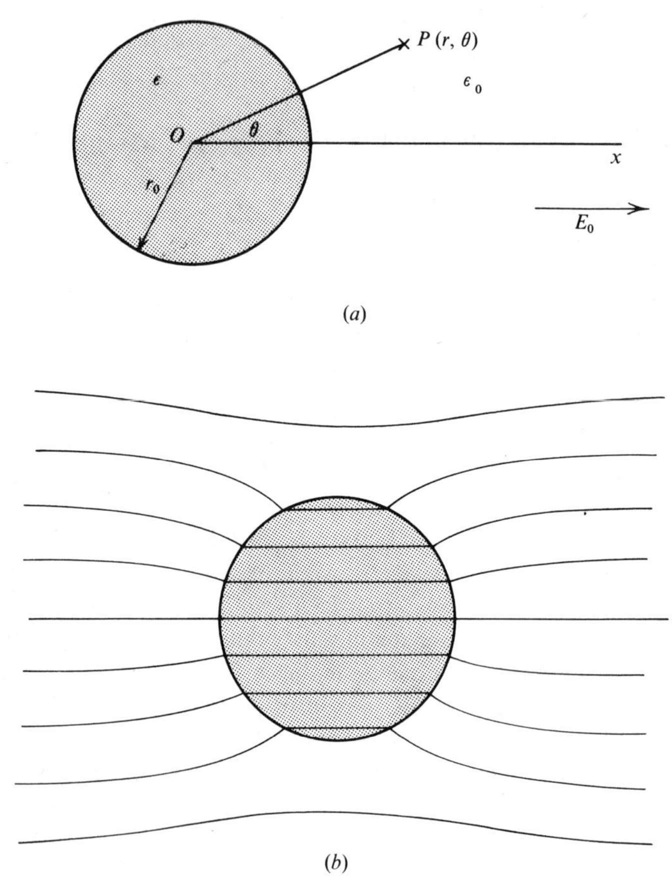


Fig. 6.3

at great distances from the dielectric. Our problem is to calculate the field inside and outside the cylinder.

We use cylindrical coordinates (fig. 6.3(*a*)) with the *z*-axis normal to the paper. Since there is no variation in the *z*-direction, the problem is a two-dimensional one and Laplace's equation reduces to

$$\frac{\partial^2 V}{\partial r^2}+\frac{1}{r}\frac{\partial V}{\partial r}+\frac{1}{r^2}\frac{\partial^2 V}{\partial \theta^2} = 0 \qquad (6.27)$$

We assume a solution of the form

$$V = R(r)\Theta(\theta)$$

where $R$ is a function of $r$ alone and $\Theta$ of $\theta$. Substituting in (6.27), we have

$$\Theta\left(\frac{\mathrm{d}^2R}{\mathrm{d}r^2}+\frac{1}{r}\frac{\mathrm{d}R}{\mathrm{d}r}\right)+\frac{R}{r^2}\frac{\mathrm{d}^2\Theta}{\mathrm{d}\theta^2} = 0$$

or
$$\frac{r^2}{R}\left(\frac{\mathrm{d}^2R}{\mathrm{d}r^2}+\frac{1}{r}\frac{\mathrm{d}R}{\mathrm{d}r}\right) = -\frac{1}{\Theta}\frac{\mathrm{d}^2\Theta}{\mathrm{d}\theta^2}$$

Since one side of this equation is independent of $\theta$, and the other of $r$, each side must be equal to the same constant, which we put equal to $n^2$. We then have the two ordinary differential equations

$$\frac{\mathrm{d}^2\Theta}{\mathrm{d}\theta^2}+n^2\Theta = 0$$

and
$$r^2\frac{\mathrm{d}^2R}{\mathrm{d}r^2}+r\frac{\mathrm{d}R}{\mathrm{d}r}-n^2R = 0$$

The solutions of these equations are

$$\Theta = a\cos n\theta+b\sin n\theta \tag{6.28}$$

$$R = cr^n+dr^{n-1} \quad \text{for } 0 < n \tag{6.29}$$

$$R = c+d\ln r \quad \text{for } n = 0 \tag{6.30}$$

where $a$, $b$, $c$ and $d$ are arbitrary constants. Any product of $\Theta$ and $R$ will satisfy (6.27) and we begin by considering which of the terms in the above solutions will enable us to meet our boundary conditions.

Referring to fig. 6.3($a$) we take the initial uniform field $E_0$ to be parallel to the $x$-axis $OX$, which is also the axis from which $\theta$ is measured. We may then conclude that the overall field will be symmetrical about the plane containing $OX$ and the $z$-axis. This rules out the solution (6.30) since, if $n = 0$, $\Theta$ is a constant and (6.30) would then indicate that the disturbance caused by the dielectric had symmetry about the $z$-axis, instead of about the plane $OXZ$. Similarly, symmetry enables us to discard the term $b\sin n\theta$ in (6.28). Finally, since increasing $\theta$ by an integral multiple of $2\pi$ must leave the value of $V$ unchanged, we conclude that $n$ must be an integer. We are thus left with terms of the forms

$$A_n r^n\cos\theta \quad \text{and} \quad B_n r^{-n}\cos\theta \tag{6.31}$$

with which to satisfy the boundary conditions.

The initial constant field $E_0$ in the $x$-direction is represented in polar coordinates by $E_0\cos\theta$ so, as $r$ tends to infinity, the total field must have this value. If we take the axis of the cylinder to be at potential zero, the conditions become

$$V = 0 \quad \text{when } r = 0 \tag{6.32}$$

$$V = -E_0 r\cos\theta \quad \text{when } r = \infty \tag{6.33}$$

Comparing (6.33) with (6.31), it seems not unlikely that we can meet these conditions using only terms for which $n = 1$, and this proves to be the case. We therefore take for the potential inside the dielectric

$$V_i = Ar\cos\theta - E_0 r\cos\theta \tag{6.34}$$

which satisfies (6.32) and, for the potential outside the dielectric,

$$V_0 = \frac{B}{r}\cos\theta - E_0 r\cos\theta \tag{6.35}$$

which satisfies (6.33).

$V$ must be continuous at the surface of the cylinder, so

$$Ar_0\cos\theta - E_0 r\cos\theta = \frac{B}{r_0}\cos\theta - E_0 r\cos\theta$$

or

$$B = Ar_0^2 \tag{6.36}$$

Since we have ensured that, at all points on the surface of the cylinder, $V$ shall be the same just inside and just outside the dielectric, it follows that the tangential component of $E$ will be the same on the two sides of the boundary. We have thus satisfied one of the conditions shown in §5.5.1 to be necessary. The other condition was that the normal component of $\boldsymbol{D}$ should be constant. Differentiating (6.34) and (6.35) with respect to $r$, to get the normal components of $\boldsymbol{E}$, multiplying by the appropriate permittivities and equating for $r = r_0$, we obtain

$$\epsilon(A - \epsilon_0)\cos\theta = -\epsilon_0\left(-\frac{B}{r_0^2} - E_0\right)\cos\theta$$

Substituting from (6.36), we find

$$A = E_0(\epsilon - \epsilon_0)/(\epsilon + \epsilon_0) \tag{6.37}$$

Inserting this value in (6.34) the expression for the potential inside the dielectric becomes

$$V_i = (-2E_0\epsilon_0\, r\cos\theta)/(\epsilon + \epsilon_0) \tag{6.38}$$

and the corresponding field in a radial direction is

$$-\frac{\partial V_i}{\partial r} = (2E_0\epsilon_0\cos\theta)/(\epsilon + \epsilon_0)$$

which is equivalent to a uniform field

$$E_i = 2E_0\epsilon_0/(\epsilon + \epsilon_0) \tag{6.39}$$

parallel to the $x$-axis.

Outside the dielectric the field is no longer parallel to the $x$-axis. It has the form shown in fig. 6.3($b$) and its components can be calculated from (6.35). $E_r$ is given by $-\partial V/\partial r$ and $E_\theta$ by $-\partial V/r\,\partial\theta$.

### 6.3.2 The two-dimensional Laplace equation in cartesian coordinates

There are many practical problems in which variation of the field in one direction (say $z$) can be assumed to be zero. This would be the case, for example, with the flow of current through a conducting sheet of constant thickness, when conditions are such that the current is distributed uniformly across the thickness. Or, again, when we consider the electrostatic field between parallel conductors of constant cross-section, when the length of each conductor is large compared with the distance between them. In such cases Laplace's equation reduces to

$$\frac{\partial^2 V}{\partial x^2}+\frac{\partial^2 V}{\partial y^2} = 0 \tag{6.40}$$

and there is a special method of dealing with this equation which yields solutions to a wide variety of problems.

Let $Z$ be a complex number, so that

$$Z = x+\mathrm{j}y \tag{6.41}$$

where both $x$ and $y$ are real numbers. Next, let $W$ be some function of $Z$, so that

$$W = f(Z) = u+\mathrm{j}v \tag{6.42}$$

where $u$ and $v$ are real functions of $x$ and $y$. We shall now quote, without proof, certain results from the theory of complex variables.

The derivatives of $W$ with respect to $Z$ may or may not have a unique value at some point $Z_0$. If the value is to be unique, $u$ and $v$ and their first partial derivatives with respect to $x$ and $y$ must be continuous in the neighbourhood of $Z_0$ and, in addition, $u$ and $v$ must satisfy the equations

$$\frac{\partial u}{\partial x} = \frac{\partial v}{\partial y} \quad \text{and} \quad \frac{\partial u}{\partial y} = -\frac{\partial v}{\partial x} \tag{6.43}$$

which are known as the Cauchy–Riemann conditions. A function of $Z$ which satisfies the above conditions is said to be *analytic* or *regular*. Most of the simple functions of $Z$ (algebraic, trigonometric, hyperbolic, exponential) are analytic except at certain points and, for our present purpose, it is safe to assume that the equations of (6.43) are satisfied. Differentiating the first of these equations with respect to $x$ and the second with respect to $y$ it becomes clear that

$$\frac{\partial^2 u}{\partial x^2}+\frac{\partial^2 u}{\partial y^2} = 0 \tag{6.44}$$

Similarly, if the first is differentiated with respect to $y$ and the second with respect to $x$,

$$\frac{\partial^2 v}{\partial x^2}+\frac{\partial^2 v}{\partial y^2} = 0 \tag{6.45}$$

Thus, both $u$ and $v$ are functions of $x$ and $y$ which satisfy Laplace's equation. They are termed *conjugate functions.*

If we multiply together the equations of (6.43), we obtain

$$\left(\frac{\partial u}{\partial x}\right)\left(\frac{\partial u}{\partial y}\right) = -\left(\frac{\partial v}{\partial x}\right)\left(\frac{\partial v}{\partial y}\right) \tag{6.46}$$

and it is shown in books on the theory of complex variables that (6.46) is precisely the condition that the two sets of curves

$$u = \text{constant} \tag{6.47}$$

$$v = \text{constant} \tag{6.48}$$

when plotted in the $xy$-plane, should intersect at right angles.

Let us now consider the application of the above theory to the solution of problems in two-dimensional electrostatic fields, where the field is produced by potentials applied to very long conductors parallel to the $z$-axis. To simplify matters we assume that there are only two such conductors, which will cut the $xy$-plane in two curves. Let us now suppose that we can find some function

$$W = f(Z) = u + \mathrm{j}v \tag{6.49}$$

such that, for two suitably chosen values of the constant in (6.47), this equation gives the curves in which the electrodes cut the $xy$-plane. Then, since (6.47) satisfies Laplace's equation, we can obtain the equations of as many equipotentials as we wish, by inserting appropriate values for the constant in (6.47). Moreover, since the curves of (6.48) cut those of (6.47) orthogonally, (6.48) must represent a set of lines of flux. The two equations (6.47) and (6.48) thus provide a complete solution of the field problem. Clearly, they also provide the solution of some different field problem in which (6.48) gives the equipotentials and (6.47) the lines of flux.

In the above discussion we have not indicated how the appropriate function

$$W = f(Z)$$

is to be found, and this is by no means always straightforward, even if it is possible. One obvious method of attack is to try a number of different functions and to find the field configurations to which they provide solutions. Over the years this method has been followed and a large store of solutions has resulted. There are, however, various techniques for enabling one to deal with a particular problem, when a solution is possible: in particular for dealing with boundary curves made up of straight lines. Further discussion is outside the scope of this book and we shall give only one example to illustrate the power of the method.

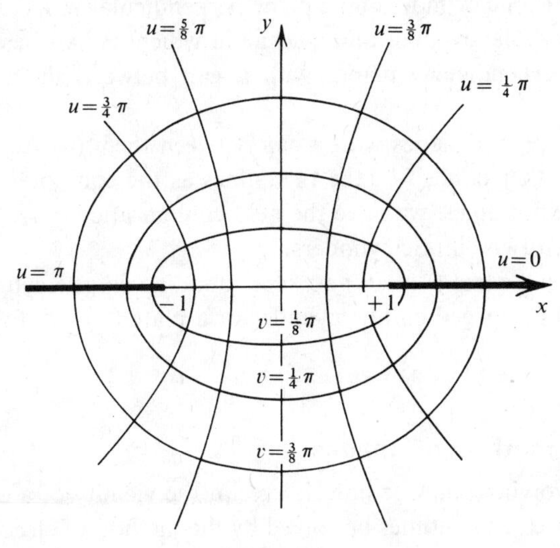


Fig. 6.4

We consider the function

$$W = f(Z) = \cos^{-1} Z$$

or
$$u + jv = \cos^{-1}(x + jy) \tag{6.50}$$

Expanding,

$$x + jy = \cos(u + jv) = \cos u \cosh v - j \sin u \sinh v$$

giving
$$x = \cos u \cosh v, \quad y = -\sin u \sinh v$$

It then follows that
$$\frac{x^2}{\cosh^2 v} + \frac{y^2}{\sinh^2 v} = 1 \tag{6.51}$$

$$\frac{x^2}{\cos^2 u} - \frac{y^2}{\sin^2 u} = 1 \tag{6.52}$$

Equation (6.51) represents a set of confocal ellipses and (6.52) a set of confocal hyperbolae which are orthogonal to the ellipses. Both sets are plotted in fig. 6.4

To appreciate what problems have been solved by the above procedure, we remember that, if the $u$-lines are taken to represent equipotentials, any two of them can be replaced by conducting cylinders of the same shapes, with an appropriate potential difference between them. The $v$-lines are then the flux lines. Thus we have formulae for equipotentials and flux lines for the following situations:

(*a*) Two hyperbolic cylinders.

(*b*) A hyperbolic cylinder and a plane perpendicular to its axis ($u = \frac{1}{2}\pi$).

(*c*) A hyperbolic cylinder and a plane in which its axis lies ($u = \pi$).

(*d*) Two perpendicular planes with a gap between them ($u = 0$ and $u = \frac{1}{2}\pi$).

(*e*) Two coplanar planes with a gap between them ($u = 0$ and $u = \pi$).

If, on the other hand, we take the $v$-lines as the equipotentials and the $u$-lines as the flux lines, we have the field configurations:

(*f*) Between two elliptic cylinders.

(*g*) Between a plate ($v = 0$) and a surrounding elliptic cylinder.

(*h*) Round a charged elliptic cylinder or a plate ($v = \infty$ for the second electrode).

### 6.3.3 The method of images

Problems involving point or line charges in the vicinity of a conductor of regular shape can sometimes be solved by the method of electrical images. By way of introduction we consider first the simple case where a point charge $Q$ is brought from infinity to a point distant $d$ from an infinite plane uncharged conductor $P$ (fig. 6.5). We stipulate that the conductor shall initially have been at zero potential, relative to a point at infinity, and that its potential shall have been kept at zero as the charge $Q$ was brought near it. The implication of these conditions will be considered later when we deal with finite conductors.

The presence of $Q$ will cause the mobile charge within the conductor to re-distribute itself in such a way as to keep $P$ at zero potential at all points. Let us enquire what would be the situation if there were no re-distribution of charge on $P$ but if, instead, there were placed on the left-hand side of $P$ a charge $-Q$ at a point distant $d$ from the plane and such that the line joining $Q$ to $-Q$ were normal to the plane. These two charges would clearly ensure that the whole of $P$ was at zero potential. Moreover, the field resulting from the two charges, which we can readily calculate, has a distribution of potential which satisfies Laplace's equation. However, we have seen (§6.2.5) that there is only one distribution of potential which will satisfy Laplace's equation and also the boundary conditions. In the present instance the only boundary condition is that $P$ should be at zero potential everywhere and this is satisfied by $Q$ and $-Q$ in the positions stated. We therefore conclude that the field to the right of $P$, which is actually caused by $Q$ and by the re-distributed charge on $P$, is exactly the same as the field that would be caused by $Q$ and $-Q$ if there were no charge on $P$. We have been able to replace the unknown distributed charge on $P$ by the simple *image charge* $-Q$ and thus to solve our problem.

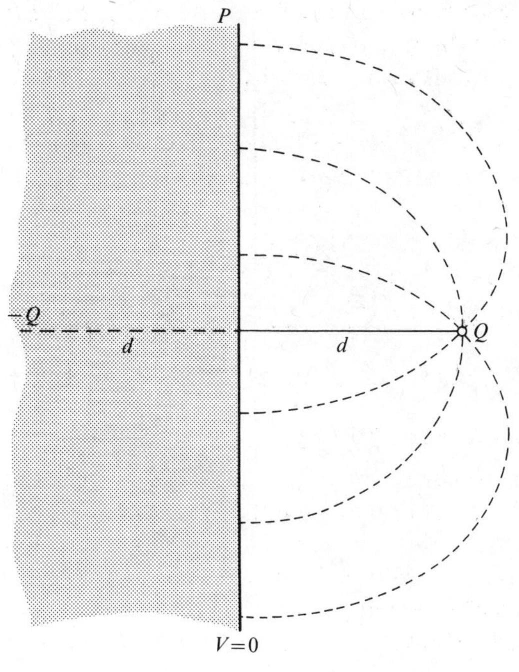


Fig. 6.5

Clearly, the application of this method depends on being able to find an appropriate image and the number of situations in which this can be done is quite limited. Nevertheless the method is often very useful and we indicate very briefly two other cases where it can be applied. In fig. 6.6(*a*), $OX$ and $OY$ are two semi-infinite conducting planes meeting in a right angle and a charge $Q$ is situated at $A$. The planes are kept at zero potential both before and after the introduction of $Q$. As before, an image charge $-Q$ at $B$ would simulate the re-distribution of charge on $OY$ if $OX$ were not present and an image charge $-Q$ at $C$ would simulate charge re-distribution on $OX$ if $OY$ were not present. Since both plates are present, we need a third image $Q$ at $D$ and it will be seen that the three images, with the original charge $Q$ at $A$, keep both plates at zero potential.

In fig. 6.6(*b*) we have a point charge $Q$ outside a conducting sphere of radius $R$. From the symmetry of the system, if an image exists, it must lie on the line joining $Q$ to the centre of the sphere $O$. Also, if we can satisfy the boundary conditions for one plane containing both $Q$ and $O$, they will be satisfied for all other such planes. As usual, we take the potential of the sphere to be zero, both before and after the introduction of $Q$.

Taking rectangular axes as shown, let $Q$ be at the point $(a, 0)$ and let us

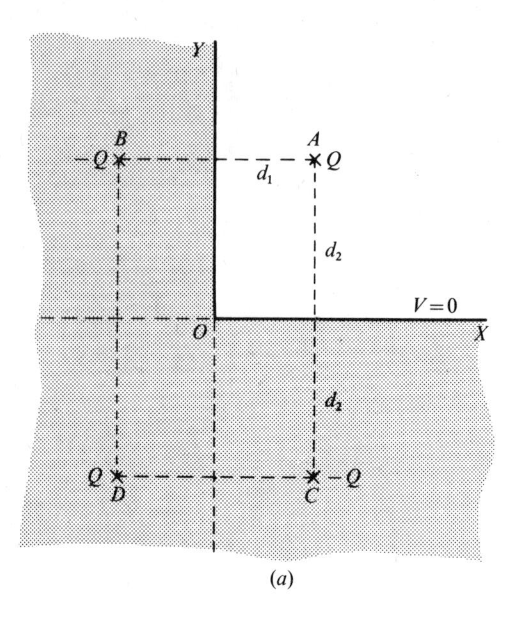


(a)

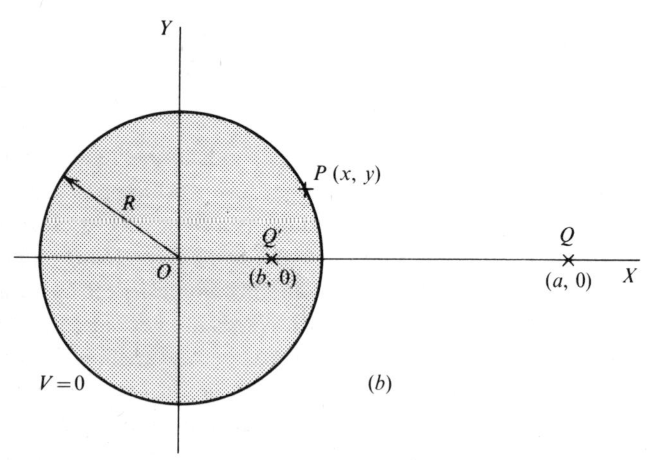


(b)

Fig. 6.6

attempt to find an image in the form of a point charge $Q'$ at the point $(b, 0)$. $P$ is any point $(x, y)$ on the circle in which the sphere cuts the plane that we are considering. Then

$$x^2+y^2 = R^2 \tag{6.53}$$

Also, if the potential at $P$ is to be zero,

$$\frac{Q}{4\pi\epsilon_0\sqrt{[(a-x)^2+y^2]}}+\frac{Q'}{4\pi\epsilon_0\sqrt{[(x-b)^2+y^2]}} = 0$$

Substituting from (6.53)

$$\frac{Q}{Q'} = -\sqrt{\frac{R^2+a^2-2ax}{R^2+b^2-2bx}} \tag{6.54}$$

If $Q'$ is to be true image, the values of $Q'$ and $b$ must be such that (6.54) is true for all values of $x$ which do not exceed $R$, and we do not yet know whether it is possible to achieve this result. We begin by choosing values of $Q'$ and $b$ which will satisfy (6.54) for the two values $x = R$ and $x = -R$. The first of these gives

$$Q/Q' = \pm(R-a)/(R-b)$$

and, for the second $\quad Q/Q' = \pm(R+a)/(R+b)$

so

$$(R-a)/(R-b) = \pm(R+a)/(R+b) \tag{6.55}$$

If we take the positive sign, we get the trivial result

$$b = a, \quad Q' = -Q$$

which is not a solution to our problem.

With the negative sign, we obtain

$$b = R^2/a, \quad Q' = -RQ/a \tag{6.56}$$

If we now substitute these values in (6.54), we find that the equation is satisfied for all values of $x$ which do not exceed $R$, so they represent a true image of the original charge $Q$.

In the three cases just considered, it has been stipulated that the conductors should be maintained throughout at zero potential relative to a point at infinity and we must now consider what this means in practice. We are interested in the field in the region of a conductor and of a charge $Q$ which is brought near it, and we must assumed that all other charges and conductors are so far away that they do not appreciably affect the field that we wish to calculate. We might, for example, suppose our system to be situated in the centre of a large room with conducting walls and the potential of these walls would then be our reference point from which other potentials are to be measured. Alternatively, and more probably, the system might be suspended above a large conducting surface, such as the earth, which would then be the reference point for potential measurements. As has been explained earlier, the earth as such has no special significance other than that it is a very large conductor from which it is often convenient to measure potential.

From the above discussion it appears that the requirement that a conductor should remain at constant zero potential when a charge $Q$ is brought near it can be satisfied if the conductor is connected to the walls of a large conducting room, or to earth, as the case may be. Such a conductor is said to be *earthed* and the three examples so far considered refer to earthed conductors. When a charge $Q$ is brought near to an earthed conductor, the potential of the latter is kept constant by the flow of charge from it to earth. This flow presumably causes some change in the potential of the earth relative to other celestial bodies, but with this we are not concerned!

We now have to investigate the alternative arrangement in which the conductor is insulated from earth, and we take as example an insulated sphere to which charge $Q$ is brought near. In this case, the total charge on the sphere must remain constant, while its potential changes as $Q$ approaches. Using the notation of fig. 6.6(*b*), an image charge $-RQ/a$, at a distance $R^2/a$ from the centre of the sphere, will take account of the re-distribution of charge on the sphere, caused by the proximity of $Q$, and will ensure that the surface of the sphere is an equipotential. However, since the sphere was initially uncharged, this image must be balanced by a positive charge $RQ/a$ on the sphere, so placed that the spherical surface remains an equipotential. Clearly, the appropriate place for this additional charge is at the centre of the sphere. Thus, in this case, the field outside the sphere is to be calculated from the three point charges; the external charge $Q$, the image charge $-RQ/a$ at distance $R^2/a$ from the centre, and the balancing charge $RQ/a$ at the centre. Finally, if the insulated sphere had carried initial charge $Q'$ before the introduction of the external charge $Q$, its effect outside the sphere would have been the same as if it had been concentrated at the centre. By the principle of superposition, the total field when $Q$ is brought into position can be calculated from three point charges: the external charge $Q$, an image charge $-RQ/a$ and a charge $(Q'+RQ/a)$ at the centre of the sphere.

We have discussed the method of images in connection with electrostatic examples, but its use is not limited to problems of this kind. We shall see later that magnetic materials exist with relative permeabilities in excess of 1000 and that a slab of such a material forms a close approximation to a magnetic equipotential region. If, therefore, a long straight filament of direct current is flowing parallel to the plane of the slab, at distance $h$ from the face, the total magnetic field can be determined by replacing the slab by an appropriate current image. In this case the image is found to be an identical current filament situated at distant $h$ behind the face of the slab, the direction of flow being the same for the real current and its image.

In suitable cases the image method can also be used with time-varying fields; for example, in calculating the total radiation from an aerial situated above a perfectly conducting plane earth.

### 6.3.4 The computer

Our discussion of the techniques available for solving Laplace's equation will have made it clear that none of the methods is of universal application and that, even when an analytical solution can be obtained, the mathematical complexity may be considerable. Solutions to a large number of Laplacian fields have been published and the reader who has a problem of this kind would be well advised to investigate what has already been done, before embarking on research of his own.†

For those fields which do not yield to analytical treatment, a computer may be used to obtain a solution. Satisfactory programs are now available, but a discussion of these is outside the scope of this book.

### 6.3.5 The method of curvilinear squares

We shall now consider an approximate method of determining fields that satisfy Laplace's equation, which depends on the free-hand plotting of equipotentials and flux lines. It is applicable only to two-dimensional problems, where there is no variation of the field in a direction normal to the plane of the paper on which the plotting is to be carried out. We shall describe the method in the first instance, by referring to the flow of current through a sheet of material of arbitrary shape, of resistivity $\rho$ and of uniform thickness $t$. It will be assumed that the current flow is parallel to the faces of the sheet and that, at any point of a face, the current density is constant across the thickness. The problem is then two-dimensional and the field configuration can be completely specified by flux lines and equipotentials drawn on a plane sheet of paper. We suppose the current to enter and leave the sheet through electrodes of high conductivity which are soldered to the sheet. These electrodes can be regarded as equipotentials and they provide some of the boundary conditions. The remaining conditions arise from the fact that no current can flow outside the sheet.

We know that equipotentials and flux lines must cut each other at right angles. Moreover, there is no limit to the number of lines of either kind that can be drawn so that, if these numbers are properly chosen, the two sets of lines will sub-divide the conductor into areas which are approximately square. They will not be perfectly square because both equipotentials and flux lines are generally curved. We shall call the areas *curvilinear squares*, and, if enough lines of both sorts have been drawn, the curvature of the sides of the areas will be so slight that it can be neglected.

† A useful collection of solutions is to be found in *Analysis and computation of electric and magnetic field problems* by K. J. Binns and P. J. Lawrenson (Pergamon Press). This book also gives references to other sources of solutions.

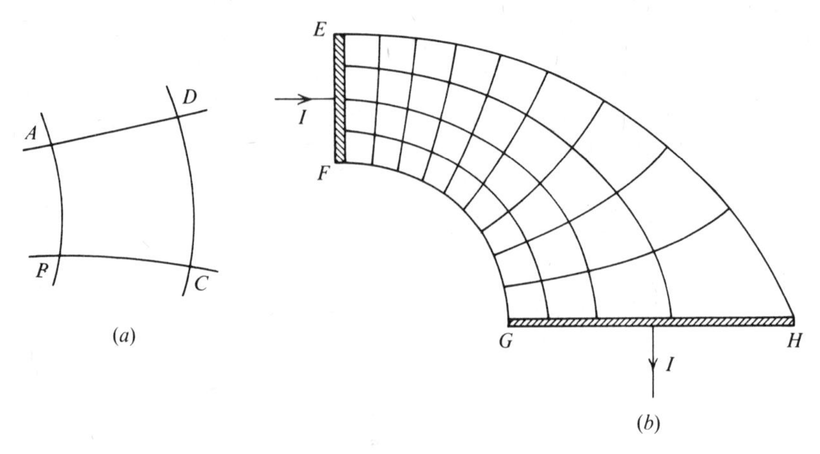


Fig. 6.7

Let us suppose that, by a procedure to be discussed later, a plot of curvilinear squares has been prepared for a particular conductor, and let us consider the flow of current across one square, such as $ABCD$ in fig. 6.7(*a*). $AB$ and $DC$ are portions of equipotentials, while $AD$ and $BC$ are parts of flux lines. Let $l$ be the length of side of the square; its resistance is then

$$R_0 = \rho l/tl = \rho/t \tag{6.57}$$

Since $R_0$ is independent of $l$, it follows that each square in a complete plot makes the same contribution to the total resistance between the electrodes.

A plot of the kind we are considering is shown in fig. 6.7(*b*), where current enters through a highly conducting electrode $EF$, flows through the irregularly shaped sheet and leaves through the electrode $GH$. $EF$ and $GH$ are equipotentials and, since no current can flow across the other boundaries, $EH$ and $FG$ are limiting lines of flow. So far as resistance is concerned, squares between two adjacent flux lines are clearly in series, while those between two adjacent equipotentials are in parallel. Since, in fig. 6.7(*b*), there are four squares between equipotentials and nine squares between flow lines, the total resistance between the electrodes is

$$9R_0/4 = 9\rho/4t$$

The reader may feel that a difficulty arises with the 'square' adjacent to the point $H$. The conductor has an acute angle here, and no skill in plotting can make this area a proper square. The answer is two-fold; first that this is only one area out of thirty-six, so its irregular shape is unlikely to cause much error in the total resistance. Second, that we can reduce the error to any desired extent by sub-dividing the plot into smaller squares.

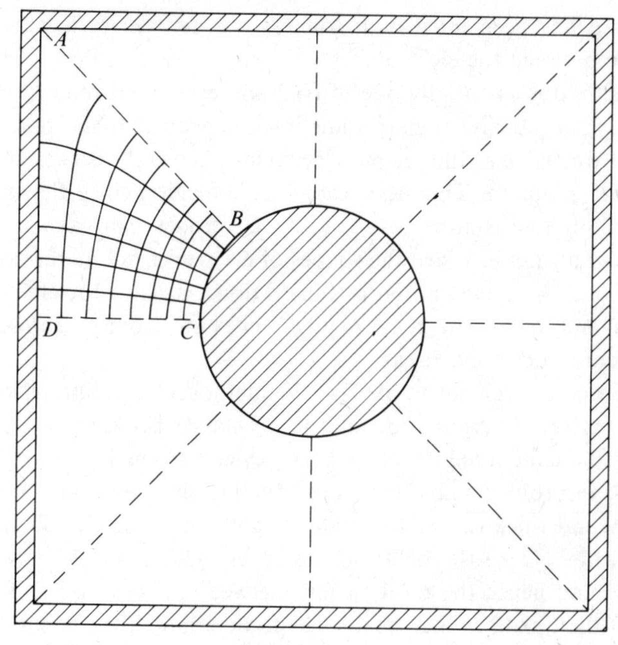


Fig. 6.8

Having learnt how to derive quantitative results from a given plot, we must next consider how the plot itself is to be produced. Suppose our problem is that indicated in fig. 6.8, where a square conducting sheet has electrodes soldered to it in the forms of the two shaded areas. We first look for possible symmetries and note that the system is symmetrical about each of the four dashed lines. We need therefore deal only with one octant, such as $ABCD$, knowing that the pattern will be repeated in the other octants. Flux lines must spread out as we pass from $BC$ to $AD$, which means that the electric field is stronger near $BC$ than near $AD$. We may thus surmise that current flow near $BC$ will not be greatly affected by the shape of the external electrode, and we begin by taking five equally spaced flux lines between $B$ and $C$. Since $BC$ and $AD$ are equipotentials, the five lines must leave $BC$ radially and meet $AD$ normally. Furthermore, the current density at the outer electrode will obviously decrease as we move from $D$ to $A$ so, as we leave $BC$, the distance between adjacent lines must increase in going from $D$ to $A$. This information is sufficient to enable us to set out, with a soft pencil in one hand and an india-rubber in the other, to sketch the five flow lines in what appear to be reasonable positions. Next we attempt to fit in the equipotentials to give curvilinear squares and this is likely to involve some modification of the flux lines. Finally, when the whole area has been filled with squares, we can derive a value for the resistance between the electrodes.

The method is essentially one of trial and error, and practice is needed to use it satisfactorily. It may sometimes happen that the plot proceeds well until we arrive at the second electrode, when there is room only for a fraction of a square. This need cause no difficulty, since the method of calculating the resistance between the electrodes can take account of fractions of a square. When the shapes of electrodes are more complicated than the ones we have chosen as illustrations, it helps to remember that flux lines tend to concentrate at highly convex areas of an electrode and to be less dense at concave areas.

It has been convenient to explain the method of curvilinear squares by referring to the determination of the resistance between two electrodes soldered to a conducting sheet, but the device has much wider application than this. In problems involving a conducting sheet, we may be given the resistivity and thickness of the material and the voltage applied between the electrodes. Once the field plot has been obtained (6.57) gives us the resistance and hence the total current between the electrodes. The distribution of the flux lines then enables us to determine the current density at any point on either electrode while, from the rate of change of potential, we can find the electric field strength. Finally, we are not limited to a conducting sheet; the electrodes may be long parallel rods, of uniform cross-section, immersed in a conducting medium.

When we come to apply the method to the investigation of electrostatic fields, the conditions are somewhat different, since there is nothing corresponding to the conducting medium, in the previous case, which limits the volume of space through which flux can pass. In principle, long parallel conducting electrodes, immersed in a medium whose relative permittivity was very large, would provide an analogous case. However, in practice, such media invariably have permittivities which vary with field strength and this invalidates the method. We thus tend to be limited to situations of two types; either we have long parallel electrodes in free space, or the field is limited because one electrode encloses the other. In the first of these the field extends without limit and, unless we make a rather large plot, it may be difficult to estimate the total flux. We shall therefore confine attention to cases, such as that represented in fig. 6.8, where one electrode surrounds the other. The electrodes are assumed to be infinitely long and we deal with unit length of the system. There is a potential difference between the electrodes.

In fig. 6.9 let $HKLM$ be one of the curvilinear squares of fig. 6.8. Let it have length of side $l$ and let the flux lines be parallel to $HK$ and $ML$. If $D$

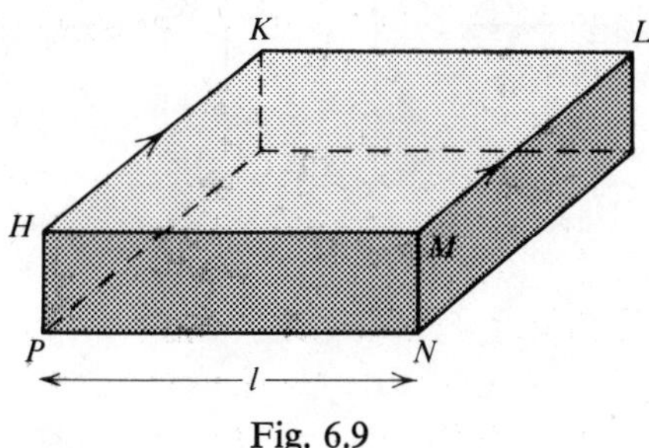


Fig. 6.9

is the value of the displacement flux density in the square, the total flux passing through $HMNP$ which, for unit length of the system has area $l$, is $Dl$. This flux corresponds to charges $+Q$ and $-Q$ respectively, which reside on those portions of the electrodes where the flux begins and ends. Thus

$$D = Q/l \tag{6.58}$$

and, with the number of squares drawn in fig. 6.8, the total charge on each electrode is $48Q$. If $\epsilon$ is the permittivity of the medium, the field strength $E$ is

$$E = D/\epsilon = Q/l\epsilon \tag{6.59}$$

and the potential difference between the opposite sides of the square, $KL$ and $HM$, is

$$V = Q/\epsilon \tag{6.60}$$

Again, with the number of squares drawn in fig. 6.8, the total potential difference between the electrodes is

$$V_0 = 8V = 8Q/\epsilon \tag{6.61}$$

If we are given $V_0$, the above equations, together with the field plot, enable us to determine $\boldsymbol{E}$ at any part of the field or the charge density on any part of either electrode. Moreover, the capacitance $C$ per unit length of the system is, from (6.61),

$$C = 48Q/(8Q/\epsilon) = 6\epsilon \text{ per metre} \tag{6.62}$$

### 6.3.6 Magnetic fields

In the foregoing discussion of methods of solving Laplace's equation we have confined our attention to electrostatic fields and to the flow of current in an extended medium. When we turn to magnetic fields, we encounter a difficulty because, in general, we do not know what boundary conditions the solution must satisfy. If the field results from the flow of current in coils situated in free space (or in air), we have no equipotential surfaces, at

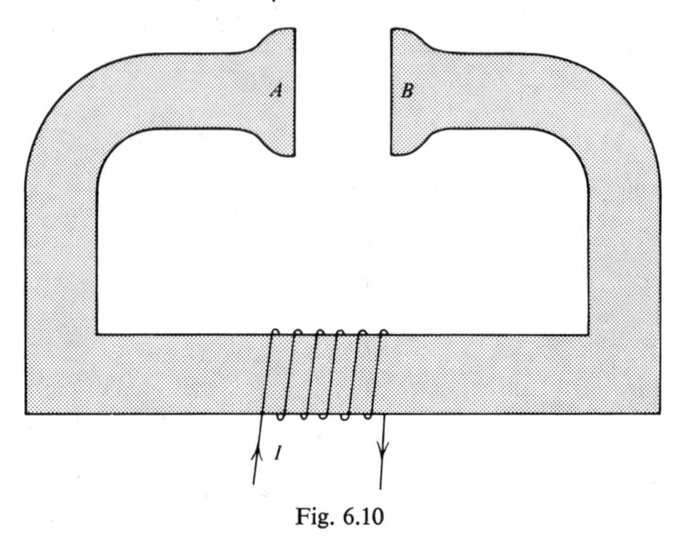


Fig. 6.10

known magnetic potentials, corresponding to the electrodes by means of which an electrostatic field is established, and Laplace's equation then gives us little help. We shall however consider two special cases where the equation can usefully be applied.

In most practical devices which make use of a magnetic field an iron core is included to concentrate the flux where it is needed. We then have to deal with a situation such as that illustrated by the electromagnet of fig. 6.10, where the flux is produced by current flowing through coils wound on an iron core, but is used in the air gap between the pole pieces $A$ and $B$. We shall be discussing problems of this kind in chapter 7 and shall then show that, to a high degree of approximation, $A$ and $B$ can be regarded as magnetic equipotential surfaces. We shall also show how the potential difference between them can be calculated, so the data needed for the application of Laplace's equation to the field in the air gap is available to us.

A second case arises when we have a system of currents which produces in free space (or air) a magnetic field with symmetry about an axis, and when we can calculate the field strength at points along this axis. As an illustration we consider the field produced by current $I$ flowing in a circular loop of wire, the diameter of the wire being negligible (fig. 6.11). Let $O$ be the centre of the coil and $r_0$ its radius. We take the $z$-axis through $O$, at right angles to the plane of the coil, and denote the position of any point $P$, distance $r$ from the axis, by its coordinates $(r, z)$.

For any point $Q$ $(0, z)$ on the axis, the solid angle subtended by the coil can readily be shown to be

$$\Omega = 2\pi[1-z/\sqrt{(z^2+a^2)}]$$

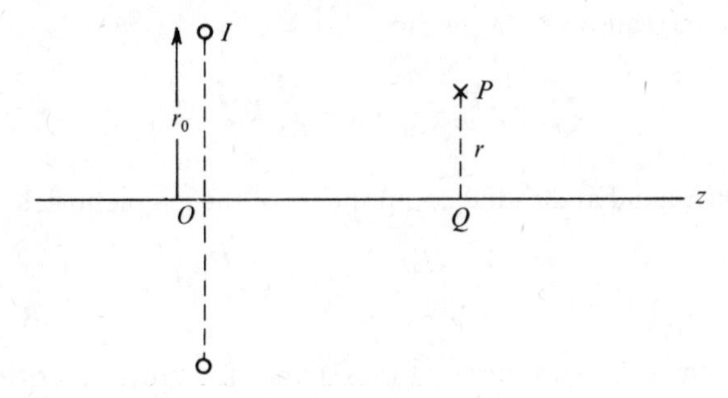


Fig. 6.11

and, as we have previously shown (§4.1.9), the magnetic potential at $Q$ will be given by

$$U = I\Omega/4\pi = \tfrac{1}{2}I[1-z/\sqrt{(z^2+a^2)}] \tag{6.63}$$

For regions where no current is flowing $U$ must obey Laplace's equation

$$\frac{\partial^2 U}{\partial r^2}+\frac{1}{r}\frac{\partial U}{\partial r}+\frac{\partial^2 U}{\partial z^2} = 0 \tag{6.64}$$

We attempt to find a solution of this equation by writing

$$U = A_0+A_2r^2+A_4r^4+\ldots \tag{6.65}$$

where $A_0$, $A_2$, ... are functions of $z$ only. Odd terms of the series are omitted since changing the sign of $r$ must not alter $U$. Then

$$\frac{1}{r}\frac{\partial U}{\partial r} = 0+2A_2+4A_4r^2+\ldots$$

$$\frac{\partial^2 U}{\partial r^2} = 0+2A_2+12A_4r^2+\ldots$$

$$\frac{\partial^2 U}{\partial z^2} = A_0''+A_2''r^2+\ldots$$

where primes denote differentiation with respect to $z$. Substituting these values in (6.64) and equating successive powers of $r$ to zero,

$$A_0''+4A_2 = 0$$

$$A_2''+16A_4 = 0 \quad \text{etc.}$$

or

$$A_2 = -\tfrac{1}{4}A_0''$$

$$A_4 = -\tfrac{1}{16}A_2'' = \tfrac{1}{64}A_0^{\text{iv}}$$

Furthermore, $A_0$ is the potential distribution $U_{0,z}$ along the $z$-axis, which is given by (6.63), and we may write

$$U_{rz} = U_{0z} - \frac{r^2}{4} U''_{0z} + \frac{r^4}{64} U^{\text{iv}}_{0z} \ldots \tag{6.66}$$

To obtain the axial and radial components of the magnetic field strength, we have

$$H_z = -\frac{\partial U_{rz}}{\partial z}, \quad H_r = -\frac{\partial U_{rz}}{\partial r} \tag{6.67}$$

Carrying out the differentiations of (6.63) and, for brevity, writing

$$p = \sqrt{(a^2+z^2)}$$

we find

$$H_z = \frac{Ia^2}{2p^3}\left[1 + \frac{3}{4}\frac{r^2}{p^4}(a^2-4z^2) + \frac{45}{64}\frac{r^4}{p^8}(a^4-12a^2z^2+8z^4)\ldots\right] \tag{6.68}$$

$$H_r = \frac{3Ia^2rz}{4p^5}\left[1 + \frac{5}{8}\frac{r^2}{p^4}(3a^2-4z^2) + \frac{15}{192}\frac{r^4}{p^8}(35a^4-140a^2z^2+56z^4)\ldots\right] \tag{6.69}$$

These series converge only when $r$ is less than $a$. In particular, when $r$ is equal to $a$, $H_z$ becomes infinite in the plane of the loop, as is to be expected since we have assumed the diameter of the wire to be negligibly small. When $r$ is greater than $a$, other series for the components of $H$ can be found by a different method, but we shall not pursue this matter.

The above discussion brings out very clearly the extent to which Laplace's equation limits the form of the magnetic field which it is possible to establish in a system with symmetry about an axis. Once the variation of the field along the axis has been settled, the value of $\boldsymbol{H}$ at all other points is determined.

### 6.3.7 Experimental methods

We have seen that Laplace's equation governs the distribution of potential in three quite different types of problem: the flow of current in a conductor, the electrostatic field and, in regions where no current flows, the magnetic field. It follows that if, with a given set of boundary conditions, we can determine the potential distribution for one type of problem, the solution will be valid for the other two types so long as the same boundary conditions apply.

We have already seen (§2.1.3) that, when current flows through a conducting sheet of uniform thickness, plots of the potential distribution can readily be obtained experimentally and this technique can be used to solve two-dimensional potential problems. If the conducting sheet is replaced by a tank of weakly conducting liquids (tap water is often satisfactory), the method can be applied, in principle, to the solution of three-dimensional problems. However, the task of measuring and recording potentials at a large number of points in three dimensions becomes excessively laborious and the method is rarely used with electrodes of arbitrary shape.

When the electrode system has symmetry about an axis, the problem is greatly simplified and the technique then becomes useful. With this symmetry, it is only necessary to measure potentials in a single plane through the axis, since the potential distribution in all similar planes will be the same. Moreover, the distribution will be unaffected if, instead of employing a complete model of the system, we use any portion bounded by two planes passing through the axis, so long as the boundary planes are insulators. Arising from these considerations there are two different experimental procedures.

In fig. 6.12(*a*), semi-cylindrical models of the system are used, with the boundary plane lying in the surface of the liquid. The models are supported from a frame (not shown) which lies above the surface of the liquid and which therefore does not disturb the field. This arrangement is the better one when accurate results are required. In the simpler scheme, the plane insulating base of the tank is tilted at an angle of about fifteen degrees from the horizontal and the liquid is only sufficient to cover a portion of this base. The straight line in which the liquid surface intersects the base is to be the axis of symmetry and only a narrow wedge-shaped sector of the system need be modelled. For most purposes the curvature of the model in planes at right angles to the axis can be neglected, so the model can be constructed of metal strips, bent to appropriate shapes. It is thus very easy to change the shapes and this is particularly convenient when the problem is to find a system which will give a desired potential distribution.

Whichever procedure is used, potentials are determined by means of a movable probe, which just touches the liquid surface (fig. 6.12(*b*)). Readings are taken from the calibrated potentiometer $P$, when balance has been attained, and are expressed as fractions of the total voltage applied between the electrodes. To avoid polarization effects at liquid–metal interfaces, it is convenient to use alternating square-wave voltages at a frequency of about 1 kHz, with a cathode-ray oscilloscope as a detector. Polarization effects cause spikes to appear on the waveform, but these die away in a fraction of a cycle and balance can readily be detected (fig. 6.12(*c*)).

For accurate work the above experimental method, and a rather similar one employing a resistance network, have been largely supplanted by the

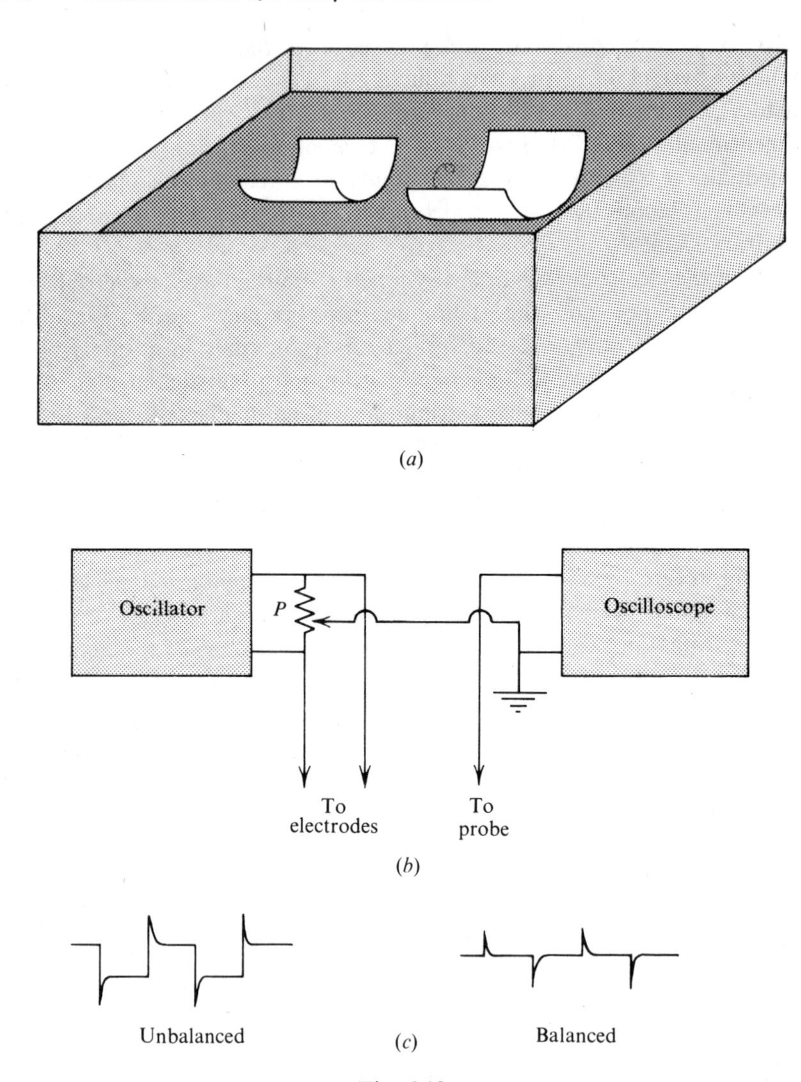


Fig. 6.12

computer, but the electrolytic tank still has its uses when one wishes to get a rough idea of the shapes of electrodes needed to produce a particular potential distribution.

## 6.4 Poisson's equation

### 6.4.1 A general solution of Poisson's equation

As we have seen, Laplace's equation enables us to determine the form of an electrostatic field when that field is produced by known potentials applied to electrodes of prescribed shape; that is, when the boundary conditions are given in terms of the potential. If, in addition, electric charge is present in the space with which we are concerned, we need Poisson's equation

$$\frac{\partial^2 V}{\partial x^2}+\frac{\partial^2 V}{\partial y^2}+\frac{\partial^2 V}{\partial z^2} = -\frac{\rho}{\epsilon_r \epsilon_0} \tag{6.70}$$

which was derived in §6.2.1. To simplify the present discussion we will postulate that the whole of the space with which we are concerned is filled with a single dielectric of relative permittivity $\epsilon_r$. When solved, (6.70) would tell us the value of $V$ at any point. However, from elementary considerations, we know that the contribution which a charge $\rho\,dv$, where $dv$ is an element of volume, makes to the potential at a point distant $r$ from $dv$, is $\rho\,dv/4\pi\epsilon_r\epsilon_0 r$. Thus we may write

$$V = \frac{1}{4\pi\epsilon_r\epsilon_0}\int_v \frac{\rho}{r}\,dv \tag{6.71}$$

where the integration must include all nett charge, whether in space or on electrodes. It follows that (6.71) is a solution of (6.70). We shall not have occasion to use this solution in connection with electrostatic fields, but shall need it later (§6.7.2).

### 6.4.2 Applications of Poisson's equation

We give below the two applications of Poisson's equation which are of considerable importance, but we shall not discuss the physical theory on which they are based. The calculations will not be used elsewhere in this book.

In fig. 6.13(*a*) $A$ is a plate which, when heated, can emit thermionic electrons. $B$ is a parallel plate, distant $d$ from $A$, and $B$ is maintained at a positive potential $V_0$ with respect to $A$. Both plates are assumed to be infinite in extent, so that edge effects can be neglected, and the space between them is evacuated.

When $A$ is cold, the potential will vary linearly as we pass from $A$ to $B$ (curve (i) of fig. 6.13(*b*)). As $A$ is heated, electrons will be emitted and will be accelerated from $A$ to $B$. Their presence in the space between the electrodes will distort the field and this effect will be greatest in the vicinity of $A$, where the electrons are moving most slowly and their density is

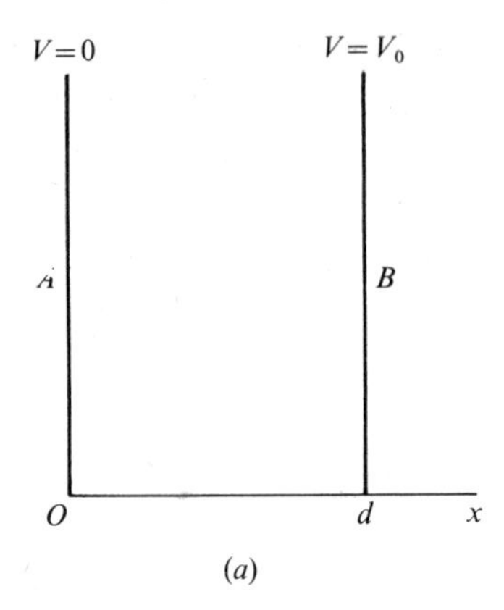


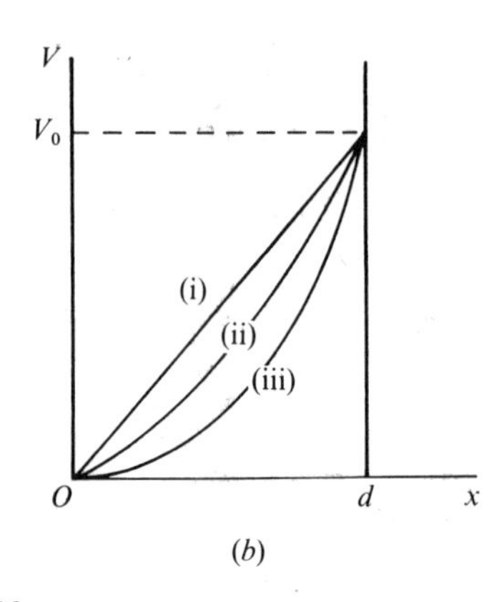


Fig. 6.13

greatest (curve (ii)). As the temperature of $A$ rises still further, a stage will be reached (curve (iii)) where the curve of $V$ against $x$ has zero slope at the origin. The electric field at the surface of $A$ is then zero. If we make the simplifying assumption that electrons are emitted from $A$ with negligible velocity (a reasonable approximation if $V_0$ is greater than, say, 20 volts), curve (iii) represents the condition when the maximum possible electron current is passing between $A$ and $B$: any increase in the current would produce a retarding electric field at the surface of $A$ and emission would cease until the field had been reduced to zero again. Thus, so long as the temperature of the emitter $A$ is sufficiently high, the current flowing to $B$ is limited by *space charge* and we wish to obtain an expression for this maximum current.

For the one-dimensional system that we are considering, Poisson's equation becomes

$$\frac{d^2V}{dx^2} = -\frac{\rho}{\epsilon_0} \tag{6.72}$$

Let $e$ and $m$ be the negative charge and mass of an electron respectively, $J$ the current density, $\rho$ the density of charge and $v$ the velocity of an electron at any point distant $x$ from $A$, where the potential is $V$. $J$ and $\rho$ are both negative and $J$ must be the same for all values of $x$ if there is to be no build-up of charge. Also

$$J = \rho v \tag{6.73}$$

and

$$\tfrac{1}{2}mv^2 = Ve \tag{6.74}$$

Substituting these values in (6.72), we find

$$\frac{d^2V}{dx^2} = -\frac{J}{\epsilon_0}\sqrt{\frac{m}{2eV}} \tag{6.75}$$

This equation can be solved by multiplying both sides by $2\mathrm{d}V/\mathrm{d}x$ and integrating. Remembering that we have taken $\mathrm{d}V/\mathrm{d}x$ to be zero at $A$, where $V$ also is zero, we have

$$\left(\frac{\mathrm{d}V}{\mathrm{d}x}\right)^2 = -\frac{4J}{\epsilon_0}\sqrt{\frac{mV}{2e}} \tag{6.76}$$

Taking the square root of both sides, integrating and putting $V = 0$ when $x = 0$, we obtain

$$V^{\frac{3}{4}} = \frac{3}{4}\left[\sqrt{\left(-\frac{4J}{\epsilon_0}\sqrt{\frac{m}{2e}}\right)}\right]x \tag{6.77}$$

Rearranging and putting $x = d$, we finally get for the current density reaching the collector

$$J = -\frac{4\epsilon_0}{9}\left(\sqrt{\frac{2e}{m}}\right)\frac{V^{\frac{3}{2}}}{d^2} \tag{6.78}$$

the negative sign meaning only that a negative current is flowing in the positive direction of $x$.

As a second example of the use of Poisson's equation we shall calculate the width of the depletion layer in a semiconductor *p–n* junction, in terms of the total potential difference across this junction. We assume the junction to be plane, with $N_a$ acceptors per unit volume on the *p*-side and $N_d$ donors per unit volume on the *n*-side. Edge effects will be neglected and we consider the case when no external voltage is applied to the junction.

A somewhat idealized representation of the equilibrium state is represented in fig. 6.14, where distances are measured along an $x$-axis at right angles to the plane of the junction, with the origin in this plane. Electrons have diffused from the *n*-side to the *p*-side to combine with holes, and holes have diffused in the reverse direction to combine with electrons. As a result of these processes, donors on the *n*-side up to some plane at $x_2$ and acceptors on the *p*-side up to some plane at $-x_1$ are left un-neutralized and this space charge produces a potential difference between the two sides which prevents further diffusion. The material to the right of $x_2$ and to the left of $-x_1$ contains mobile carriers and so cannot sustain an electric field. Thus one condition for equilibrium is that the distances $x_1$ and $x_2$ must adjust themselves so that the flux of $\boldsymbol{D}$ originating on the un-neutralized donors, all ends on the un-neutralized acceptors. Hence

$$N_a x_1 = N_d x_2 \tag{6.79}$$

Between $x = 0$ and $x = -x$, Poisson's equation gives us

$$\frac{\mathrm{d}^2V}{\mathrm{d}x^2} = \frac{N_a e}{\epsilon_r \epsilon_0} \tag{6.80}$$

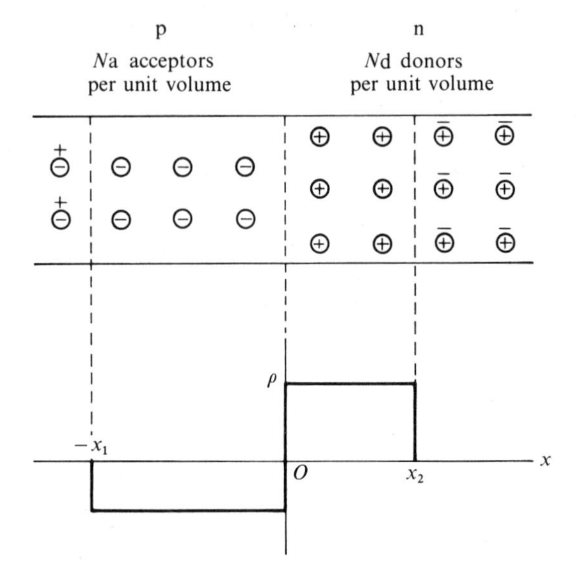


Fig. 6.14

where $\epsilon_r$ is the relative permittivity of the material and $e$ is the electronic charge, which is equal to the charge on each un-neutralized acceptor. Then

$$\frac{dV}{dx} = \frac{N_a ex}{\epsilon_r \epsilon_0} + C_1$$

But, as we have seen, the electric field is zero when $x = -x_1$, and then $dV/dx = 0$. Thus

$$\frac{dV}{dx} = \frac{N_a e}{\epsilon_r \epsilon_0}(x+x_1)$$

and
$$V = \frac{N_a e}{2\epsilon_r \epsilon_0}(x+x_1)^2 + C_2$$

If we take the zero of potential to be that when $x = 0$,

$$C_2 = -N_a e x_1^2/2\epsilon_r \epsilon_0$$

and
$$V_{-x1} = -N_a e x_1^2/2\epsilon_r \epsilon_0 \tag{6.81}$$

Similarly
$$V_{x2} = N_d e x_1^2/2\epsilon_r \epsilon_0 \tag{6.82}$$

If $V_0$ is the total potential difference between the two sides,

$$V_0 = V_{x2} - V_{-x1} = e(N_a x_1^2 + N_d x_2^2)/2\epsilon_r \epsilon_0$$

Substituting from (6.79),

$$x_1 = [2\epsilon_r \epsilon_0 N_d V_0 / e N_a (N_a + N_d)]^{\frac{1}{2}}$$

and $$x_2 = [2\epsilon_r \epsilon_0 N_a V_0 / eN_d(N_a+N_d)]^{\frac{1}{2}}$$

Hence the width of the depletion layer is

$$d_0 = x_1+x_2 = [2\epsilon_r \epsilon_0 V_0(N_a+N_d)/eN_a N_d]^{\frac{1}{2}} \qquad (6.83)$$

The theory of semiconductors indicates how $V_0$, $N_a$ and $N_d$ can be determined, but we are not concerned with these problems. We must, however, consider whether the use of the macroscopic value of the relative permittivity $\epsilon_r$ in the above calculation is justifiable. When appropriate values of the various quantities are inserted in (6.83) it appears that $d_0$ is unlikely to be less than $10^{-6}$ cm and will usually be greater. Thus $d_0$ will generally be at least one hundred times the distance between adjacent atoms in the semiconductor and the use of the macroscopic value for $\epsilon_r$ is reasonable.

## 6.5 Differential forms of the magnetic field equations

### 6.5.1 Ampère's law

In chapter 4 (§4.1.10) we expressed Ampère's law in the form

$$\oint_l \boldsymbol{H}\cdot d\boldsymbol{l} = I \qquad (6.84)$$

We now wish to obtain the corresponding differential equation relating to the values of the quantities at a point. Since $I$ is not a point function, we re-write (6.84) as

$$\oint_l \boldsymbol{H}\cdot d\boldsymbol{l} = \oint_S \boldsymbol{J}\cdot\boldsymbol{n}dS \qquad (6.85)$$

where $\boldsymbol{J}$ is the current density and $S$ is a surface bounded by the closed path $l$.

In fig. 6.15 let the current density at the origin be $J$, with components $J_x$, $J_y$ and $J_z$, and consider the rectangular path $OYPZ$, with sides of length $dy$ and $dz$ respectively. The current passing through this rectangle is $J_x dy dz$. Let $\boldsymbol{H}$ be the magnetic field strength at the origin, with components $H_x$, $H_y$ and $H_z$. The field along $OY$ is $H_y$, while that along $ZP$ is

$$H_y + \frac{\partial H_y}{\partial z} dz$$

and similarly for the fields along $OZ$ and $YP$. Hence (6.85) for the loop $OYPZ$ becomes

$$H_y dy - \left(H_y + \frac{\partial H_y}{\partial z} dz\right) dy + \left(H_z + \frac{\partial H_z}{\partial y} dy\right) dz - H_z dz = J_x dy dz$$

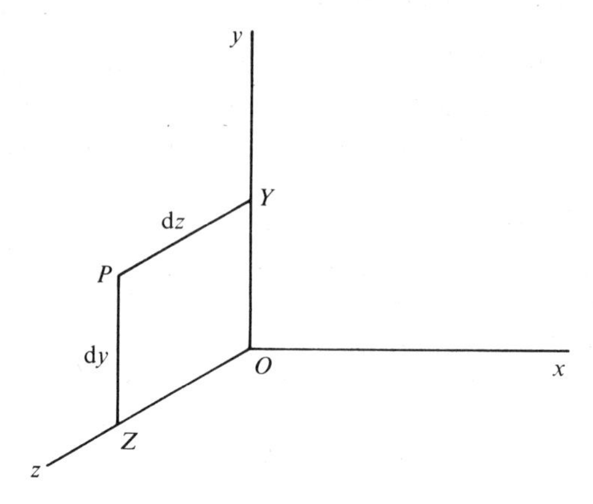


Fig. 6.15

or
$$\frac{\partial H_z}{\partial y} - \frac{\partial H_y}{\partial z} = J_x \tag{6.86}$$

Similarly
$$\frac{\partial H_x}{\partial z} - \frac{\partial H_z}{\partial x} = J_y \tag{6.87}$$

and
$$\frac{\partial H_y}{\partial x} - \frac{\partial H_x}{\partial y} = J_z \tag{6.88}$$

These three equations can be written as a single vector equation, in the form

$$\left(\frac{\partial H_z}{\partial y} - \frac{\partial H_y}{\partial z}\right) \boldsymbol{i} + \left(\frac{\partial H_x}{\partial z} - \frac{\partial H_z}{\partial x}\right) \boldsymbol{j} + \left(\frac{\partial H_y}{\partial x} - \frac{\partial H_x}{\partial y}\right) \mathbf{k}$$
$$= J_x \boldsymbol{i} + J_y \boldsymbol{j} + J_z \boldsymbol{k} = \boldsymbol{J} \tag{6.89}$$

The left-hand side of this equation is a function which frequently occurs in analytical work and is given the name curl $\boldsymbol{H}$ (or sometimes rot $\boldsymbol{H}$). Thus we have

$$\text{curl}\, \boldsymbol{H} = \boldsymbol{J} \tag{6.90}$$

In the above derivation it has been tacitly assumed that $\boldsymbol{J}$ is a conduction current density. We shall see later (§9.2) that, for a complete description, it is necessary to add another term, representing the displacement current, to the right-hand side of (6.90).

### 6.5.2 Electromagnetic induction

From the foregoing, it will be clear that the operation of taking the curl of a vector at any point is equivalent to taking the line integral of the vector round an infinitesimally small loop surrounding the point, and dividing the result by the area of the loop. We now recall that we have earlier been concerned with line integrals of the vector $\boldsymbol{E}$ and we therefore enquire whether it would be profitable to apply the results of the previous section to these integrals.

An electric field can be caused either by the presence of charges or by a changing magnetic field and, for any closed loop, we may take account of these two components by writing

$$\oint_l \boldsymbol{E}\cdot d\boldsymbol{l} = \oint_l \boldsymbol{E}_e\cdot d\boldsymbol{l} + \oint_l \boldsymbol{E}_m\cdot d\boldsymbol{l} \tag{6.91}$$

For the electrostatic field $E_e$, resulting from charges, we know that

$$\oint \boldsymbol{E}_e\cdot d\boldsymbol{l} = 0 \tag{6.92}$$

Also, we know that

$$\oint_l \boldsymbol{E}_m\cdot d\boldsymbol{l} = -\frac{d\phi}{dt} = -\frac{d}{dt}\oint_S \boldsymbol{B}\cdot\boldsymbol{n}\,dS$$

where $S$ is a surface bounded by the loop and $\phi$ is the magnetic flux passing through the loop. Hence (6.91) becomes

$$\oint_l \boldsymbol{E}\cdot d\boldsymbol{l} = -\frac{d}{dt}\oint_S \boldsymbol{B}\cdot\boldsymbol{n}\,dS \tag{6.93}$$

If the loop $l$ and the surface $S$ are stationary and the change in $\phi$ results entirely from the change in $\boldsymbol{B}$, we may write (6.93) as

$$\oint_l \boldsymbol{E}\cdot d\boldsymbol{l} = -\oint_S \frac{\partial \boldsymbol{B}}{\partial t}\cdot\boldsymbol{n}\,dS \tag{6.94}$$

and, in the differential form, this becomes

$$\text{curl}\,\boldsymbol{E} = -\frac{\partial \boldsymbol{B}}{\partial t} \tag{6.95}$$

Applications of this equation and of (6.90) will be given later.

## 6.6 Vector formulae

### 6.6.1 The del operator

We have seen that when any of the terms grad, div or curl precedes a quantity, scalar or vector as the case may be, it indicates that certain differential operations are to be performed on that quantity. Grad, div and curl are therefore referred to as *operators*. Purely as a matter of notation, it is convenient to introduce another operator which we term *del* and write $\nabla$. In cartesian coordinates it is defined by the identity

$$\text{del} \equiv \nabla \equiv \boldsymbol{i}\frac{\partial}{\partial x}+\boldsymbol{j}\frac{\partial}{\partial y}+\boldsymbol{k}\frac{\partial}{\partial z} \tag{6.96}$$

Similar definitions can be formulated for other coordinate systems, but we shall not need to consider these. We shall now show how grad, div and curl can be expressed in terms of del.

First, we let del operate on a scalar quantity $V$. We then have

$$\text{del } V = \nabla V = \boldsymbol{i}\frac{\partial V}{\partial x}+\boldsymbol{j}\frac{\partial V}{\partial y}+\boldsymbol{k}\frac{\partial V}{\partial z} = \text{grad } V \tag{6.97}$$

Similarly, we may let del operate on a vector, but we must then decide whether the scalar or the vector product is to be taken. Taking the scalar product with a vector such as $\boldsymbol{E}$,

$$\nabla\cdot\boldsymbol{E} = \left(\boldsymbol{i}\frac{\partial}{\partial x}+\boldsymbol{j}\frac{\partial}{\partial y}+\boldsymbol{k}\frac{\partial}{\partial z}\right)(\boldsymbol{i}E_x+\boldsymbol{j}E_y+\boldsymbol{k}E_z)$$

$$= \frac{\partial E_x}{\partial x}+\frac{\partial E_y}{\partial y}+\frac{\partial E_z}{\partial z} = \text{div } \boldsymbol{E} \tag{6.98}$$

On the other hand, if we take the vector product,

$$\nabla\times\boldsymbol{E} = \left(\boldsymbol{i}\frac{\partial}{\partial x}+\boldsymbol{j}\frac{\partial}{\partial y}+\boldsymbol{k}\frac{\partial}{\partial z}\right)\times(\boldsymbol{i}E_x+\boldsymbol{j}E_y+\boldsymbol{k}E_z)$$

and, on multiplying out the right-hand side of this question, we readily find that

$$\nabla\times\boldsymbol{E} = \text{curl } \boldsymbol{E} \tag{6.99}$$

### 6.6.2 Double application of the ∇ operator

If $V$ is any vector, $\nabla V$, the gradient of $V$, is a vector and we may wish to find its divergence, which we can denote by $\nabla\cdot(\nabla V)$. Since

$$\nabla\cdot(\nabla V) = (\nabla\cdot\nabla)V = \nabla^2 V$$

div grad $V$ is usually denoted by $\nabla^2 V$. In component form this becomes

$$\nabla^2 V = \frac{\partial^2 V}{\partial x^2}+\frac{\partial^2 V}{\partial y^2}+\frac{\partial^2 V}{\partial z^2} \tag{6.100}$$

and Laplace's equation may be expressed in the form

$$\nabla^2 V = 0 \tag{6.101}$$

For this reason the operator

$$\nabla^2 = \frac{\partial^2}{\partial x^2}+\frac{\partial^2}{\partial y^2}+\frac{\partial^2}{\partial z^2}$$

is generally called the *Laplacian* operator.

It is found to be convenient to make use of the expression $(\nabla\cdot\nabla)\boldsymbol{E}$ in which $\nabla^2$ operates on a *vector* quantity $\boldsymbol{E}$, although we have not so far attached any meaning to such an operation. We therefore adopt *as a definition*

$$(\nabla\cdot\nabla)\boldsymbol{E} = \nabla^2\boldsymbol{E} = \frac{\partial^2\boldsymbol{E}}{\partial x^2}+\frac{\partial^2\boldsymbol{E}}{\partial y^2}+\frac{\partial^2\boldsymbol{E}}{\partial z^2} \tag{6.102}$$

or, in component form,

$$\nabla^2\boldsymbol{E} = \boldsymbol{i}\nabla^2E_x+\boldsymbol{j}\nabla^2E_y+\boldsymbol{k}\nabla^2E_z \tag{6.103}$$

This extension of the meaning of $\nabla^2$ is made because it provides a useful shorthand notation.

Another double operation of $\nabla$ that we shall encounter is the gradient of the divergence of a vector $\boldsymbol{E}$. In component form, this immediately gives

$$\text{grad div } \boldsymbol{E} = \nabla(\nabla\cdot\boldsymbol{E}) = \boldsymbol{i}\left(\frac{\partial^2E_x}{\partial x^2}+\frac{\partial^2E_y}{\partial x\,\partial y}+\frac{\partial^2E_z}{\partial x\,\partial z}\right)$$

$$+\boldsymbol{j}\left(\frac{\partial^2E_x}{\partial x\,\partial y}+\frac{\partial^2E_y}{\partial y^2}+\frac{\partial^2E_z}{\partial y\,\partial z}\right)+\boldsymbol{k}\left(\frac{\partial^2E_x}{\partial x\,\partial y}+\frac{\partial^2E_y}{\partial y\,\partial z}+\frac{\partial^2E_z}{\partial z^2}\right) \tag{6.104}$$

Other double operators will be considered in the next section.

### 6.6.3 Vector identities

So far, we have considered the operation of $\nabla$ on a single quantity, but we often wish to know the results when it operates on the sum or product of two quantities. It is therefore convenient to have a table of these results for future reference. Similarly, the results of the double application of $\nabla$ can, in some cases, be expressed in useful alternative forms. In table 6.1 we give a list of the more important vector identities that arise in these ways. In the table $\boldsymbol{E}$ and $\boldsymbol{F}$ are vectors, while $U$ and $V$ are scalars.

In each case the proof of an identity can be obtained by carrying out the operations indicated on the two sides, after expressing any vectors in terms of their rectangular components. We shall give two examples, leaving the remainder as exercises for the reader.

(*a*) *To prove that*

$$\nabla\times(\boldsymbol{E}\times\boldsymbol{F}) \equiv \boldsymbol{E}\nabla\cdot\boldsymbol{F}-\boldsymbol{F}\nabla\cdot\boldsymbol{E}+(\boldsymbol{F}\cdot\nabla)\boldsymbol{E}-(\boldsymbol{E}\cdot\nabla)\boldsymbol{F} \tag{6.105}$$

We need only show that the $x$-components are the same on the two sides, since similar results will be obtained for the $y$- and $z$-components.

For the left-hand side,

$$\boldsymbol{E}\times\boldsymbol{F} = \boldsymbol{i}(E_yF_z-E_zF_y)+\boldsymbol{j}(E_zF_x-E_xF_z)+\boldsymbol{k}(E_xF_y-E_yF_x)$$

Table 6.1

| |
|---|
| $\nabla(U+V) \equiv \nabla U+\nabla V$ |
| $\nabla\cdot(\boldsymbol{E}+\boldsymbol{F}) \equiv \nabla\cdot\boldsymbol{E}+\nabla\cdot\boldsymbol{F}$ |
| $\nabla\times(\boldsymbol{E}+\boldsymbol{F}) \equiv \nabla\times\boldsymbol{E}+\nabla\times\boldsymbol{F}$ |
| $\nabla(UV) \equiv U\nabla V+V\nabla U$ |
| $\nabla\cdot(V\boldsymbol{E}) \equiv \boldsymbol{E}\cdot\nabla V+V\nabla\cdot\boldsymbol{E}$ |
| $\nabla\cdot(\boldsymbol{E}\times\boldsymbol{F}) \equiv \boldsymbol{F}\cdot\nabla\times\boldsymbol{E}-\boldsymbol{E}\cdot\nabla\times\boldsymbol{F}$ |
| $\nabla\times(V\boldsymbol{E}) \equiv \nabla V\times\boldsymbol{E}+V\nabla\times\boldsymbol{E}$ |
| $\nabla\times(\boldsymbol{E}\times\boldsymbol{F}) \equiv \boldsymbol{E}\nabla\cdot\boldsymbol{F}-\boldsymbol{F}\nabla\cdot\boldsymbol{E}+(\boldsymbol{F}\cdot\nabla)\boldsymbol{E}-(\boldsymbol{E}\cdot\nabla)\boldsymbol{F}$ |
| $\nabla\cdot\nabla V \equiv \nabla^2 V$ |
| $\nabla\cdot\nabla\times\boldsymbol{E} \equiv 0$ |
| $\nabla\times\nabla V \equiv 0$ |
| $\nabla\times\nabla\times\boldsymbol{E} \equiv \nabla(\nabla\cdot\boldsymbol{E})-\nabla^2\boldsymbol{E}$ |

Thus

$$\nabla\times(\boldsymbol{E}\times\boldsymbol{F}) = \text{curl}\,(\boldsymbol{E}\times\boldsymbol{F})$$

$$= \boldsymbol{i}\left\{E_x\frac{\partial F_y}{\partial y}+F_y\frac{\partial E_x}{\partial y}-E_y\frac{\partial F_x}{\partial y}-F_x\frac{\partial E_y}{\partial y}\right.$$

$$\left.-E_z\frac{\partial F_x}{\partial z}-F_x\frac{\partial E_z}{\partial z}+E_x\frac{\partial F_z}{\partial z}\right\}+\boldsymbol{j}\{\ldots\}+\boldsymbol{k}\{\ldots\} \tag{6.106}$$

For the right-hand side, $\boldsymbol{F}\cdot\nabla$ is to be interpreted as the operator resulting from the scalar product of $\boldsymbol{F}$ and $\nabla$. Or

$$\boldsymbol{F}\cdot\nabla = F_x\frac{\partial}{\partial x}+F_y\frac{\partial}{\partial y}+F_z\frac{\partial}{\partial z} \tag{6.107}$$

Thus the right-hand side can be written as

$$(\boldsymbol{i}E_x+\boldsymbol{j}E_y+\boldsymbol{k}E_z)\left(\frac{\partial F_x}{\partial x}+\frac{\partial F_y}{\partial y}+\frac{\partial F_z}{\partial z}\right)-(\boldsymbol{i}F_x+\boldsymbol{j}F_y+\boldsymbol{k}F_z)\left(\frac{\partial E_x}{\partial x}+\frac{\partial E_y}{\partial y}+\frac{\partial E_z}{\partial z}\right)$$

$$+\left(F_x\frac{\partial}{\partial x}+F_y\frac{\partial}{\partial y}+F_z\frac{\partial}{\partial z}\right)(\boldsymbol{i}E_x+\boldsymbol{j}E_y+\boldsymbol{k}E_z)$$

$$-\left(E_x\frac{\partial}{\partial x}+E_y\frac{\partial}{\partial y}+E_z\frac{\partial}{\partial z}\right)(\boldsymbol{i}F_x+\boldsymbol{j}F_y+\boldsymbol{k}F_z)$$

On collecting the terms in the $x$-components it becomes apparent that these are identical with those in (6.106).

(*b*) *To prove that*

$$\nabla\times\nabla\times\boldsymbol{E} = \nabla(\nabla\cdot\boldsymbol{E})-\nabla^2\boldsymbol{E} \tag{6.108}$$

$$\nabla\times\boldsymbol{E} = \boldsymbol{i}\left(\frac{\partial E_z}{\partial y}-\frac{\partial E_y}{\partial z}\right)+\boldsymbol{j}\left(\frac{\partial E_x}{\partial z}-\frac{\partial E_z}{\partial x}\right)+\boldsymbol{k}\left(\frac{\partial E_y}{\partial x}-\frac{\partial E_x}{\partial y}\right)$$

Hence, the $x$-component of the left-hand side is

$$\boldsymbol{i}\left(\frac{\partial^2 E_y}{\partial x\,\partial y}-\frac{\partial^2 E_x}{\partial y^2}-\frac{\partial^2 E_x}{\partial z^2}+\frac{\partial^2 E_z}{\partial x\,\partial z}\right)$$

and, from (6.103) and (6.104), this is identical with the $x$-component of the right-hand side. Similarly, for the $y$- and $z$-components.

## 6.7 The magnetic vector potential

### 6.7.1 Definition of the vector potential

In this section we shall quote, without proof, certain results of vector analysis. It is hoped that the treatment will appear plausible but, for a complete mathematical justification, the reader must consult more advanced texts. Furthermore, for simplicity, we shall restrict our discussion to fields in free space.

In the earlier part of this book we have encountered two different types of vector field. In electrostatics we saw that the line integral of $\boldsymbol{E}$ round any closed path was always equal to zero. Thus

$$\oint \boldsymbol{E}\cdot d\boldsymbol{l} = 0$$

or, in differential form $\quad \text{curl}\,\boldsymbol{E} = \nabla\times\boldsymbol{E} = 0 \quad (6.109)$

Any field for which an equation like (6.109) holds good is said to be *irrotational*. A consequence of the equation is that the field strength $\boldsymbol{E}$ at any point can be derived from a unique scalar potential $V$.

On the other hand, the magnetic flux density $\boldsymbol{B}$ produced by current density $\boldsymbol{J}$ obeys the equation

$$\text{curl}\,\boldsymbol{B} = \nabla\times\boldsymbol{B} = \mu_0\boldsymbol{J} \tag{6.110}$$

Further properties of this field are that $\boldsymbol{B}$ is a flux vector and that lines of $\boldsymbol{B}$ form closed loops. Thus

$$\text{div}\,\boldsymbol{B} = \nabla\cdot\boldsymbol{B} = 0 \tag{6.111}$$

Fields of this type are said to be *solenoidal* and, as we have seen, $\boldsymbol{B}$ cannot be derived from a unique scalar potential. The use of the magnetic scalar potential $U$ is limited to paths which do not enclose any current.

There is a theorem due to Helmholtz which states that the most general vector field which we can imagine can be expressed as the sum of a solenoidal and an irrotational field. In other words, a vector field is completely specified by its divergence and its curl.

In view of the limitation attaching to the magnetic scalar potential $U$, we now seek a mathematical quantity which, for the flux density $\boldsymbol{B}$, will play a part similar to that performed by the potential $V$ in electrostatics. Our reason for doing this is purely mathematical, in the hope that the new quantity will help in the solution of problems. We have no reason to expect that the new quantity will have any simple physical interpretation.

Our starting point is the identity (table 6.1, §6.6.3) which states that, for any vector $\boldsymbol{A}$,

$$\text{div}\,(\text{curl}\,\boldsymbol{A}) = \boldsymbol{\nabla}\cdot\boldsymbol{\nabla}\times\boldsymbol{A} = 0 \tag{6.112}$$

Curl $\boldsymbol{A}$ thus represents a solenoidal field and, since the field of $\boldsymbol{B}$ is also solenoidal, it is permissible to write

$$\boldsymbol{B} = \text{curl}\,\boldsymbol{A} = \boldsymbol{\nabla}\times\boldsymbol{A} \tag{6.113}$$

where $\boldsymbol{A}$ is some unknown vector, whose properties have still to be determined, and which may or may not prove useful. We note that (6.113) is not sufficient to define $\boldsymbol{A}$ since if $\boldsymbol{A}'$ is any other irrotational vector,

$$\text{curl}\,(\boldsymbol{A}+\boldsymbol{A}') = \text{curl}\,\boldsymbol{A}+\text{curl}\,\boldsymbol{A}' = \text{curl}\,\boldsymbol{A} \tag{6.114}$$

and the vector $(\boldsymbol{A}+\boldsymbol{A}')$ would have satisfied (6.113) as well as $\boldsymbol{A}$. To define $\boldsymbol{A}$ uniquely, we need to specify the divergence of $\boldsymbol{A}$ and this we shall shortly do.

Substituting (6.113) in (6.110), we have

$$\text{curl}\,(\text{curl}\,\boldsymbol{A}) = \boldsymbol{\nabla}\times(\boldsymbol{\nabla}\times\boldsymbol{A}) = \mu_0\boldsymbol{J} \tag{6.115}$$

But, from (6.108),

$$\boldsymbol{\nabla}\times(\boldsymbol{\nabla}\times\boldsymbol{A}) = \boldsymbol{\nabla}(\boldsymbol{\nabla}\cdot\boldsymbol{A})-\boldsymbol{\nabla}^2\boldsymbol{A} = \mu_0\boldsymbol{J} \tag{6.116}$$

Since div $\boldsymbol{A}$ is still at our disposal, we now decide to put

$$\text{div}\,\boldsymbol{A} = \boldsymbol{\nabla}\cdot\boldsymbol{A} = 0 \tag{6.117}$$

and (6.116) becomes

$$\boldsymbol{\nabla}^2\boldsymbol{A} = -\mu_0\boldsymbol{J} \tag{6.118}$$

$\boldsymbol{A}$ is known as the *magnetic vector potential* and is defined by the equations (6.113) and (6.117).†

## 6.7.2 Properties of the magnetic vector potential

The expression

$$\text{curl}\,\boldsymbol{A} = \boldsymbol{\nabla}\times\boldsymbol{A} = \boldsymbol{B}$$

† In some textbooks the defining equation for $\boldsymbol{A}$ is taken to be curl $\boldsymbol{A} = \boldsymbol{H}$; equations containing $\boldsymbol{A}$ are then modified accordingly.

has the corresponding integral form (§6.5.1)

$$\oint_l \boldsymbol{A}\cdot d\boldsymbol{l} = \oint_S \boldsymbol{B}\cdot\boldsymbol{n}\,dS = \phi \tag{6.119}$$

where $\phi$ is the magnetic flux passing through a surface $S$ and the line integral of $\boldsymbol{A}$ is to be taken round the boundary of this surface. If $\phi$ is changing with time, the e.m.f. induced round the boundary is equal to $-d\phi/dt$. Thus, if $\boldsymbol{E}$ is the electric field at any point which is caused by the changing $\phi$,

$$\text{e.m.f.} = \oint_l \boldsymbol{E}\cdot d\boldsymbol{l} = -\frac{\partial}{\partial t}\oint_l \boldsymbol{A}\cdot d\boldsymbol{l} \tag{6.120}$$

Turning now to (6.118), we see that this equation is similar to Poisson's equation (6.7), except that vectors are involved instead of scalars. It is equivalent to three scalar equations,

$$\begin{aligned} \nabla^2 A_x &= -\mu_0 J_x \\ \nabla^2 A_y &= -\mu_0 J_y \\ \nabla^2 A_z &= -\mu_0 J_z \end{aligned} \tag{6.121}$$

Each of which is of the form of Poisson's equation. Referring to (§6.4.1) we see that one form of solution for these equations can be written

$$A_x = \frac{\mu_0}{4\pi}\int_v \frac{J_x}{r}\,dv \tag{6.122}$$

with similar expressions for $A_y$ and $A_z$. In (6.122) $J_x$ is the $x$-component of $\boldsymbol{J}$ flowing through a small element of volume $dv$, which is distant $r$ from the point at which $A_x$ is being calculated. The three equations for the components of $\boldsymbol{A}$ can be combined to give the single vector equation

$$\boldsymbol{A} = \frac{\mu_0}{4\pi}\int_v \frac{\boldsymbol{J}}{r}\,dv \tag{6.123}$$

We thus have a means of calculating the vector potential resulting from a known distribution of current.

The vector potential plays an important part in more advanced electromagnetic theory, so the reader should be aware of its meaning and properties. We give below one example of its use, but shall not otherwise be concerned with it in this book.

### 6.7.3 Neumann's expression for mutual inductance

We wish to find the mutual inductance between two loops of wire (fig. 6.16) by calculating the flux linked with loop 2 as a result of current $I_1$ in loop 1. The dimensions of the cross-section of the wire in each loop

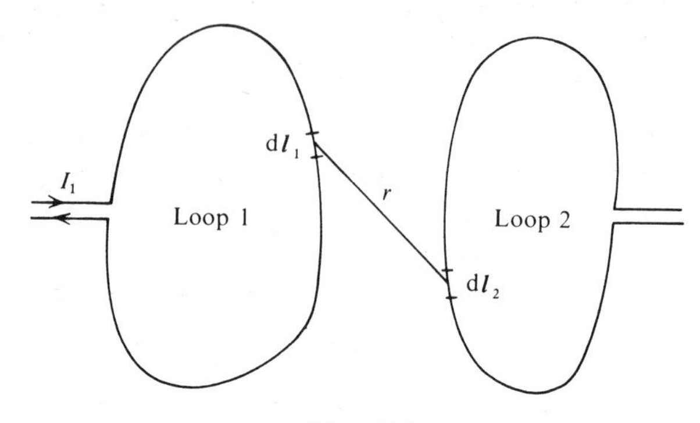


Fig. 6.16

are negligible in comparison with the dimensions of the loops and of the distance between them.

From (6.119) the flux linked with loop 2 is

$$\phi = \oint_{l_2} \boldsymbol{A}\cdot d\boldsymbol{l}_2 \tag{6.124}$$

where $\boldsymbol{A}$ is the vector potential at the element $d\boldsymbol{l}_2$. In loop 1 we have constant current $I_1$ flowing through wire of constant cross-section, so

$$\boldsymbol{J}dv = I_1 d\boldsymbol{l}_1$$

and (6.123) becomes

$$\boldsymbol{A} = \frac{\mu_0 I_1}{4\pi}\oint_{l_1}\frac{d\boldsymbol{l}_1}{r} \tag{6.125}$$

Using this value for $\boldsymbol{A}$ in (6.124) and putting $I_1$ equal to unity, we find

$$M = \frac{\mu_0}{4\pi}\oint_{l_2}\oint_{l_1}\frac{d\boldsymbol{l}_1\cdot d\boldsymbol{l}_2}{r} \tag{6.126}$$

This is Neumann's formula for $M$. We note that it is completely symmetrical in $l_1$ and $l_2$, so the expression would have been unchanged if we had considered the flux linked with loop 1 as a result of current in loop 2. In other words

$$M_{12} = M_{21}$$

as we assumed earlier.

## 6.8 Worked example

An uncharged conducting sphere of radius $R$ is placed in an electric field $\boldsymbol{E}$, which was previously uniform and which remains uniform at a great distance from the sphere. We may suppose the initial uniform field to be produced by equal positive and negative point charges situated a large distance apart. Make use of this fact to obtain an expression for the potential in the vicinity of the sphere, taking the potential of the sphere itself as zero.

*Solution.* Take an axis in the direction of the uniform field and let the centre of the sphere be at the origin of coordinates. We suppose the uniform field $\boldsymbol{E}$ to be produced by charge $-Q$ at distance $D$ along the positive direction of the axis, together with charge $+Q$ at distance $-D$. Each of these charges produces at the origin a field $Q/4\pi\epsilon_0 D^2$, so

$$E = Q/2\pi\epsilon_0 D^2 \tag{6.127}$$

To obtain a truly uniform field, we suppose both $Q$ and $D$ to increase without limit, but $Q/D^2$ remains finite in accordance with (6.127).

We have seen (§6.3.3) that, for all points outside the conducting sphere, the effect of re-distribution of charge on the sphere can be taken into account by replacing the sphere by two images of the charges $+Q$ and $-Q$. These images will be $-RQ/D$ at the point $-R^2/D$ on the axis and $+RQ/D$ at the point $+R^2/D$. The distance $2R^2/D$ between these images decreases as $D$ increases and, in the limit, they form a perfect dipole of moment

$$p = \frac{RQ}{D}\frac{2R^2}{D} = 2R^3Q/D^2$$

Substituting for $R^2/D$ from (6.127) we find

$$p = 2R^3\, 2\pi\epsilon_0 E = 4\pi\epsilon_0 R^3 E$$

The potential $V$ in the vicinity of the sphere will be the sum of two components; one arising from the dipole and the other from the uniform field.

At any point $(r, \theta)$ the dipole component will be (§3.6.6)

$$V_1 = p\cos\theta/4\pi\epsilon_0 r^2 = R^3E\cos\theta/r^2$$

The component from the uniform field will be the work that must be done on unit charge to move it from the origin to $(r, \theta)$. This is

$$V_2 = -Er\cos\theta$$

Thus, for the total potential $V$, we have

$$V = V_1 + V_2 = \left(\frac{R^3}{r^2} - r\right) E\cos\theta$$

## 6.9 Problems

1. The space between the conductors of a spherical capacitor is filled with a dielectric of permittivity $\epsilon$. The conductors have radii $R$ and $3R$ respectively and a potential difference $V_0$ is maintained between them.

Use Laplace's equation in spherical coordinates to find the field strength at a point distant $2R$ from the centre of the system.

2. A very long conducting uncharged cylinder, of radius $R$, is introduced into an electric field which was previously uniform and of strength $E$. It may be assumed that the field remains uniform at great distances from the cylinder. The axis of the cylinder is at right angles to the field.

Derive expressions for the components $E_r$ and $E_\theta$ of the field in the vicinity of the cylinder.

3. Show that example 2 can be solved by the complex-variable method, using the transformation

$$W = -E\left(z-\frac{R^2}{z}\right)$$

4. A charge $Q$ resides on an insulated conducting sphere of radius $b$. Show that a charge $q$, at a distance $d$ from the centre, where $d > b$, will not always be repelled. (University of Newcastle-upon-Tyne, 1974.)

5. In the problem illustrated in fig. 6.6($a$), how would the field be altered if the semi-infinite planes $OX$ and $OY$ were insulated from earth?

6. In the problem of fig. 6.5, show that, as a result of bringing $Q$ into position, the total charge flowing from $P$ to earth is equal to $Q$.

7. Two conducting planes $A$ and $B$ meet at right angles. A wire with a charge $Q$ coulomb per unit length is held parallel to the planes, at a distance of $3a$ from plane $A$ and a distance $2a$ from plane $B$.

Using the method of images, or otherwise, determine the density of the induced charge at $P$, the foot of the perpendicular from the wire to the plane $A$. (University of London, 1973.)

8. A conducting sphere of radius $a$ carries a charge $Q$ and there is a charge $q$ at a point $A$ which is at a distance $b$ from the centre. Show that the charge density $\sigma$ at any point $P$ on the surface of the sphere is given by:

$$\sigma = \frac{-aq}{4\pi}\left(\frac{b^2}{a^2}-1\right)\frac{1}{AP^3}+\frac{(Q+aq/b)}{4\pi a^2}$$

(University of Newcastle-upon-Tyne 1974.)

9. An uncharged conducting sphere of radius $a$ is placed in an electric field $\boldsymbol{E}_0$ which was previously uniform and which remains uniform at a great distance from the sphere.

Verify that

$$V = Ar\cos\theta+\frac{B}{r^2}\cos\theta$$

is a solution of Laplace's equation in spherical coordinates for this particular case and use it to determine the components of electric field $E_r$ and $E_\theta$ in the vicinity of the sphere.

10. A figure having the form and dimensions shown in fig. 6.17 is cut from a thin sheet of resistance alloy having resistivity $\rho$ and uniform thickness $t$. Current is led into and out of the material through heavy copper bars $A$ and $B$ respectively, which are soldered to the alloy and may be assumed to have negligible resistance.

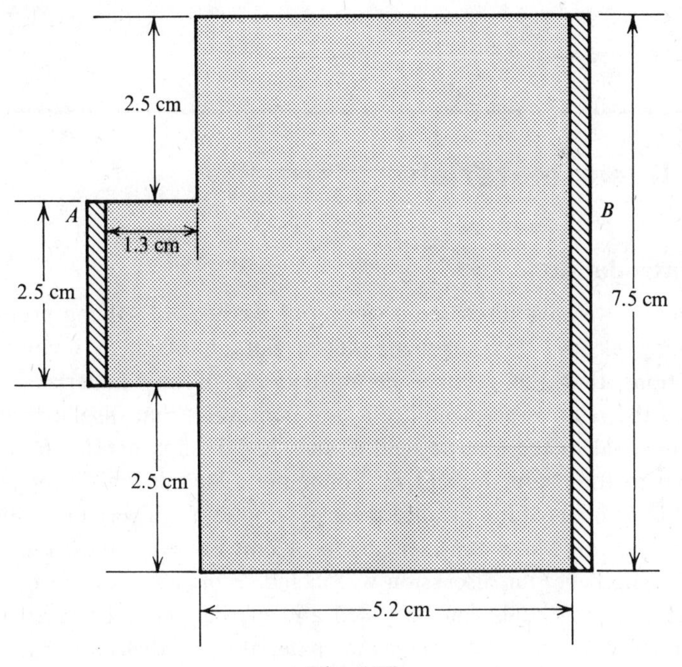


Fig. 6.17

Sketch curvilinear squares for the current flow and estimate the resistance between $A$ and $B$.

11. A point charge $Q$ is placed at a height $h$ in free space above a semi-infinite slab of dielectric of relative permittivity $\epsilon_r$. By considering the boundary conditions show that the field in the dielectric is that of a charge $2Q/(\epsilon_r+1)$ at the position of the original charge $Q$, and that the field outside the dielectric is that of the original charge $Q$ and its image $-Q(\epsilon_r-1)/(\epsilon_r+1)$ at a point depth $h$ in the dielectric. (University of Southampton, 1974.)

# 7

# Non-linear materials

## 7.1 Introduction

In chapter 6 we have discussed a variety of methods of solving problems concerning electric and magnetic fields, but all of them involve the assumptions that the relative permittivity $\epsilon_r$ of any material in the field and the relative permeability $\mu_r$ are constants; that displacement $\boldsymbol{D}$ is proportional to the electric field $\boldsymbol{E}$ and that the flux density $\boldsymbol{B}$ is proportional to the magnetic field $\boldsymbol{H}$. For many materials which are of the greatest technological importance these assumptions are very far from the truth and we must now consider how problems involving these materials are to be handled. Our discussion will include a brief account of the way in which such materials can be tested and of the results obtained from these tests. We shall treat magnetic materials and dielectric materials separately: it is possible for a single material to affect both electric and magnetic fields, but we shall not consider problems of this kind.

## 7.2 Magnetic materials

### 7.2.1 Types of magnetic material

The great majority of all materials have relative permeabilities which differ from unity by less than one part in a thousand. They may be classified into *paramagnetic* materials, for which $\mu_r$ is slightly greater than unity, and *diamagnetic* materials whose relative permeabilities are slightly less than unity. Their magnetic properties are of no interest technologically and we shall not consider them further.

A few materials have relative permeabilities very much greater than unity; almost always greater than 100 and sometimes as high as $10^6$. They are classified as ferromagnetic materials, which are metals or alloys, and ferrimagnetic materials, which have low electrical conductivities. The distinction between these two classes will be of no importance until we come to consider their properties and uses.

### 7.2.2 The form of the specimen to be tested

A number of experimental procedures have been devised for measuring particular properties of magnetic materials, but we shall be concerned with only one: the measurement of $\boldsymbol{B}$ as a function of $\boldsymbol{H}$. For this purpose we need to ensure that the whole of the specimen is subjected to a uniform field $\boldsymbol{H}$, which we can calculate, and we must then have some means of measuring $\boldsymbol{B}$. We might, for example, consider placing a rod of the material along the axis of a long solenoid which, as we have seen earlier (§4.2.3) produces a nearly uniform field in free space. However, further consideration of this arrangement shows that the presence of the specimen with high relative permeability would seriously distort the magnetic field and the ends of the specimen would not experience the same value of $\boldsymbol{H}$ as the centre portion. This end effect might be reduced to negligible proportions by choosing as specimen a cylindrical rod with length, say, one hundred times its radius, but a more satisfactory procedure is to get rid of the end effect altogether. This we can do by getting rid of the ends and using as our specimen a toroid or anchor ring, uniformly wound with an even number of layers of wire (fig. 7.1). This is still not quite an ideal arrangement since, as we have seen earlier (§4.2.5), if $N_1$ is the total number of turns in the winding, $\boldsymbol{H}$ in the toroid at a distance $r$ from its centre is given by

$$H = N_1 I/2\pi r \tag{7.1}$$

and is directed circumferentially. Thus the material corresponding to the greatest value of $r$ is subjected to a slightly lower magnetic field than that for which $r$ is least. However, if we make the diameter of the toroid large compared with the dimensions of the cross-section of the material and if we take $r$ to be the mean radius of the toroid, the field calculated from (7.1) will be sufficiently accurate for our purpose.

As we shall see later, it will be necessary to vary the current through the winding and also to reverse it. A convenient circuit for these purposes is shown in fig. 7.1, where $P$ is an electronically-stabilized constant-current power supply, whose output can be varied. $A$ is an ammeter and $S$ a reversing switch. To measure changes in the flux round the toroid, a second winding of $N_2$ turns, covering a portion of the ring, is connected to a fluxmeter, an instrument to be described in the next section.

The material to be tested may have a solid cross-section but, to reduce eddy currents (§4.4.6), magnetic cores are often built up from stampings cut from thin sheets, the stampings being electrically insulated from each other. Alternatively, cores may be fabricated from thin tape. The different methods of fabrication may cause differences in the magnetic properties

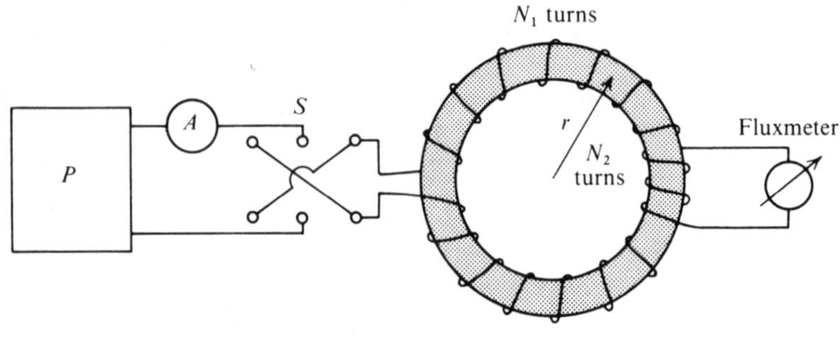


Fig. 7.1

of the material, which must therefore be tested in the form in which it is to be used.

### 7.2.3 The Grassot fluxmeter

The construction of the Grassot fluxmeter resembles that of a moving-coil galvanometer, in which the deflection of the coil is indicated either by a pointer or by a beam of light. The essential difference from a galvanometer is that the coil of a fluxmeter is wound on an insulating frame and is suspended by a fine silk fibre. Current is led to and from the coil through very thin silver strips, in such a way that the mechanical couple acting on the coil is negligibly small, whatever the position of the coil.

We suppose this instrument to be connected to the coil of $N_2$ turns in fig. 7.1. If the flux round the toroid changes, an e.m.f. will be induced in $N_2$ and this will cause a current to flow through the fluxmeter, bringing about a deflection of its coil. However, as soon as the coil begins to move in the field of the permanent magnet in the fluxmeter, a second e.m.f. will be induced in the coil and, by Lenz's law, it will be in a direction to oppose the motion of the coil. Since all mechanical couples have been eliminated, it is this second e.m.f. alone which controls the angle through which the coil is deflected.

Let $\phi$ be the flux in the toroid threading the coil $N_2$ and let this flux change from an initial value $\phi_1$ to a final value $\phi_2$. In the fluxmeter itself let the suspended coil have $N_3$ turns, each of area $S$, moving in a radial field of flux density $\boldsymbol{B}$. Then, when current $I$ flows through the coil, the torque acting on it (4.74) is $T$ where

$$T = N_3 SBI$$

and, if $J$ is the moment of inertia of the coil and $\omega$ its angular velocity,

$$N_3 SBI = J\frac{\mathrm{d}\omega}{\mathrm{d}t}$$

or
$$I = \frac{J}{N_3SB}\frac{d\omega}{dt} \tag{7.2}$$

The e.m.f. induced in the circuit as a result of the flux change in the toroid is of magnitude
$$E_1 = N_2\frac{d\phi}{dt} \tag{7.3}$$

Similarly, an opposing e.m.f. $E_2$ is induced in the circuit by the motion of the fluxmeter coil, where
$$E_2 = N_3SB\frac{d\theta}{dt} \tag{7.4}$$

If $L$ is the total inductance of the circuit and $R$ its resistance, the equation for current in the circuit is

$$N_2\frac{d\phi}{dt} = L\frac{dI}{dt}+RI+N_3SB\frac{d\theta}{dt} \tag{7.5}$$

and, substituting from (7.2), this becomes

$$N_2\frac{d\phi}{dt} = L\frac{dI}{dt}+\frac{J}{N_3SB}\frac{d\omega}{dt}+N_3SB\frac{d\theta}{dt} \tag{7.6}$$

We now integrate this equation over an interval from $t_1$ to $t_2$ to give

$$N_2(\phi_2-\phi_1) = L(I_2-I_1)+\frac{J}{N_3SB}(\omega_2-\omega_1)+N_3SB(\theta_2-\theta_1)$$

If we take $t_1$ to be an instant before any of the changes had started, and $t_2$ to be a later time when the toroid flux had reached its final value $\phi_2$, the motion of the fluxmeter coil had ceased and $I$ had decayed to zero, we have

$$I_1 = I_2 = \omega_1 = \omega_2 = 0$$

and
$$N_2(\phi_2-\phi_1) = N_3SB(\theta_2-\theta_1) \tag{7.7}$$

Thus the change in $\theta$ is proportional to the change in flux-turns linked with the circuit connected to the fluxmeter. The simplest method of finding the constant of proportionality $N_3SB$ is to calibrate the instrument with air-cored coils in which the changes in flux-turns can be calculated. For example, one might use the circuit of fig. 7.1, with an air-cored toroid replacing the iron-cored toroid.

The fluxmeter provides an interesting example of the operation of Lenz's law since examination of (7.7) shows that the change in flux linkage caused by change in $\phi$ is exactly counterbalanced by the change in flux linkage resulting from the motion of the coil; the total flux linkage remains constant.

We have described the fluxmeter with reference to our particular problem of measuring flux changes in a toroid which is being tested, but the instrument has much wider application. It is usually supplied with one or more 'search coils' in the form of flat coils, each with a known number of turns of known area. To measure an unknown magnetic field, the coil is placed with its plane at right angles to the field. A reading of the fluxmeter is taken and the change of reading noted when the search coil is suddenly withdrawn to a place where the flux density is negligibly small. The flux density of the field can then be calculated. If the direction of the field is not known, it can be found by varying the position of the search coil, while watching the fluxmeter, until the maximum flux is linked with the coil.

The sensitivity of a search coil can be reduced by shunting it with a resistance. It is left as an exercise for the reader to show that, if $R_1$ is the resistance of the coil and $R_2$ that of the shunt, the sensitivity is reduced in the ratio $R_2/(R_1+R_2)$. The ratio is independent of the resistance of the fluxmeter and of the inductance of the circuit.

The foregoing treatment suggests that the calibration of a fluxmeter is independent of the resistance of the circuit, but this cannot be true when the resistance increases without limit. The apparent paradox arises because, in (7.5), we have assumed that the whole damping of the coil arises from induced current. In fact there will always be some damping as a result of the motion of the coil through air. When the resistance of the circuit is large enough the air damping will become comparable with the electromagnetic damping and our theory will no longer be valid. As a rule air damping will not be important so long as the circuit resistance does not exceed about ten times the fluxmeter resistance.

### 7.2.4 Methods of test and general results

We wish to obtain a $B$–$H$ curve for a specimen of material in the form of a toroid, using the circuit of fig. 7.1. Using (7.1) there is no difficulty in setting $H$ to any pre-determined value. The fluxmeter will enable us to measure changes in the total flux $\phi$ round the toroid and therefore, knowing the number of turns $N_2$ and the area of cross-section $A$ of the material, to deduce changes in the flux density $B$, from the equation

$$B_2 - B_1 = (\phi_2 - \phi_1)/AN_2 \tag{7.8}$$

However, we have no means of finding either $B_2$ or $B_1$ directly; only their difference. It is therefore essential, when carrying out a test, to start with the material in some definite known state, to which it can be returned at will. One obvious choice is to begin with the toroid in an unmagnetized

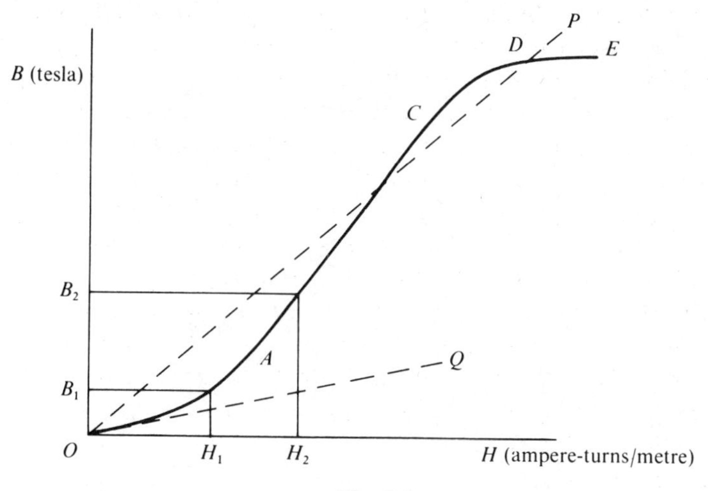


Fig. 7.2

state, since this state can easily be produced by passing alternating current through the winding of the toroid and gradually reducing this current to zero. Starting from this unmagnetized state we can now switch on a magnetizing current corresponding to a field strength $H_1$ (fig. 7.2) and, from the resulting deflection of the fluxmeter, deduce the value $B_1$ of the flux density. We now demagnetize the sample again and repeat the process for a new value $H_2$ of $H$. In this way a complete *initial magnetization curve* $OACDE$ can be built up. Such a curve normally consists of four fairly well defined regions. Between $O$ and $A$ the slope is increasing rapidly, from $A$ to $C$ the curve is nearly straight, between $C$ and $D$ the slope decreases until, between $D$ and $E$ the curve is once more straight and nearly parallel to the axis of $H$. In this final state the material is said to be saturated; it is no longer contributing to an increase in flux density.

It is clear that the permeability, defined as the ratio of $B$ to $H$, varies markedly with $H$. Two values are sometimes of interest. If $OQ$ is a tangent to the curve at the origin, we may derive from its slope the *initial permeability*. A second, rather indefinite, average permeability can be derived from the slope of a line such as $OP$. This is often useful for rough calculations since the material will not generally be used under conditions which cause it to become saturated. For magnetic fields greater than that corresponding to the point $C$, a large increase in magnetizing current produces only a small increase in $B$ and is uneconomic. Furthermore, in this region, the permeability falls quite rapidly.

A second method of testing magnetic materials arises from the fact that such materials are widely used in alternating-current equipment where the

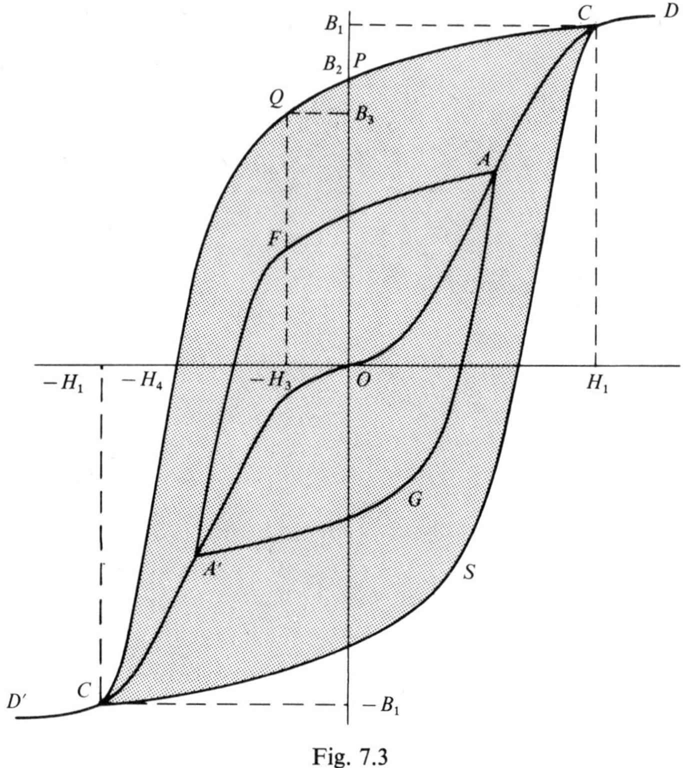


Fig. 7.3

flux is reversed many times a second. We shall shortly see that the behaviour of a material, when the magnetizing field is changed, depends very much on its initial state. However, if the material is subjected to a sufficiently strong field $H_1$ and that field is then reversed a number of times – say ten – the material gets into what may be termed a cyclic state; it has, so to speak, lost all memory of its initial state. Suppose that, for material in this state, the field is reversed from $H_1$ to $-H_1$. Then, if the flux density corresponding to $H_1$ were $B_1$, symmetry ensures that the flux density for $-H_1$ will be $-B_1$. Thus the reversal has caused a total change of $2B_1$, which can be determined from the fluxmeter reading. By taking a series of values of $H$ and putting the material into a cyclic state for each, we can find the corresponding values of $B$ and so build up a *reversal curve*, as shown in fig. 7.3 as $DCAOA'C'D'$ which is symmetrical about the origin.

The portion of this curve which lies in the first quadrant is similar to that of fig. 7.2 and it might be thought that the two should be identical, since reversal of the field necessarily causes the material to pass through an unmagnetized state. However, we shall shortly see that reversal does not cause the material to pass through the state $B = 0$, $H = 0$, so the conditions are not the same as for the curve of fig. 7.2 and there is no reason why the two curves should be identical.

From the reversal curve it is a simple matter to find out what happens to the material during a reversal. Suppose, for example, we start from point $C$ in fig. 7.3, having reached this point by repeated reversals of field $H_1$. If we now reduce $H$ to zero, we shall obtain a fluxmeter deflection $(B_1 - B_2)$ corresponding to a change to some point such as $P$. Returning to $C$, by repeated reversals of $H_1$, we now reduce to $-H_3$, getting a fluxmeter deflection corresponding to $(B_1 - B_3)$, to give another point $Q$ on the curve. Continuing in this way the curve $CQC'$ can be obtained and the opposite half $C'SC$ then drawn to give symmetry about the origin.

A curve such as $CPQC'SC$ is known as a hysteresis loop or curve. $AFA'GA$ is a similar loop for a different initial value of the magnetizing field and there is no limit to the number of such curves that can be drawn. The portion of the loop lying in the second quadrant is known as the *demagnetization curve*. For any loop such as $CPQC'SC$, the flux density $B_2$ which remains after $H$ has been reduced to zero is known as the *remanent flux density*. The magnetic field $-H_4$ needed to reduce the flux density to zero is called the *coercive force*. Clearly the values of remanent flux density and coercive force depend on the particular loop under consideration. When they refer to a loop whose tips have been taken well into the saturation region, the remanent flux density is known as the *retentivity* (denoted by $B_R$), while the coercive force is termed the *coercivity* (denoted by $H_C$).

To end this brief account of some of the ways in which magnetic materials are tested, we consider a situation which arises in certain electronic components, particularly transformers, where a magnetizing winding carries a small alternating component of current superimposed on a much larger direct current. The conditions are illustrated in fig. 7.4, where the direct current produces field $H_1$ and brings the material to the point $P$ on the magnetization curve. The alternating component of current causes the field to vary between $H_1 - \Delta H$ and $H_1 + \Delta H$ and, at first sight, we might suppose this to cause the material to vary between $Q$ and $R$ on the curve. What actually happens is that an increase in $H$, from any point on the initial magnetization curve, causes the state point to move up this curve in the usual way. On the other hand a decrease in $H$, followed by subsequent return to the initial value, causes the state point to move reversibly round a small subsidiary hysteresis loop. Thus after a few cycles of the alternating component of $H$, the state point will be traversing a loop such as

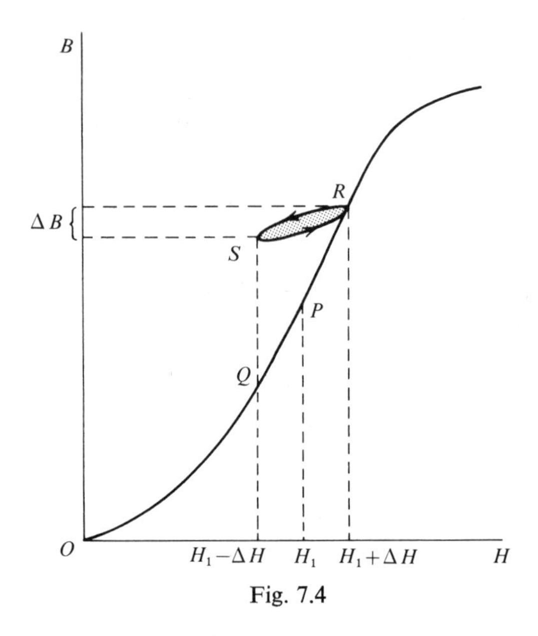


Fig. 7.4

*RS*. The mean slope of this loop, $\Delta B/2\Delta H$, represents the effective permeability of the material for the alternating component of the field. It is termed the *incremental permeability* and may not be more than about one-tenth of the slope of the initial magnetization curve.

While all ferromagnetic and ferrimagnetic materials behave in the manner discussed above, there is wide variation from one material to another in quantities such as relative permeability, maximum flux density, coercivity, retentivity and so forth, and we shall return to this matter later (§7.2.6). One final point should be made. It is never safe to predict how a magnetic material will behave unless tests have been carried out under the conditions in which the material is to be used.

### 7.2.5 Hysteresis loss

We have shown that, when a material is taken round a cycle of magnetization, returning to its initial state, its representative point on a $B$–$H$ diagram traverses a hysteresis loop. It will now be proved that the area enclosed by this loop is related to an energy loss in the material.

As before we consider a toroidal specimen of mean radius $r$ and cross-sectional area $A$, which is sufficiently small for both $B$ and $H$ to be considered uniform throughout the material. The specimen is magnetized by current $I$ flowing through a uniform winding of $N_1$ turns, so that the field $H$ is given by

$$H = N_1 I/2\pi r \tag{7.9}$$

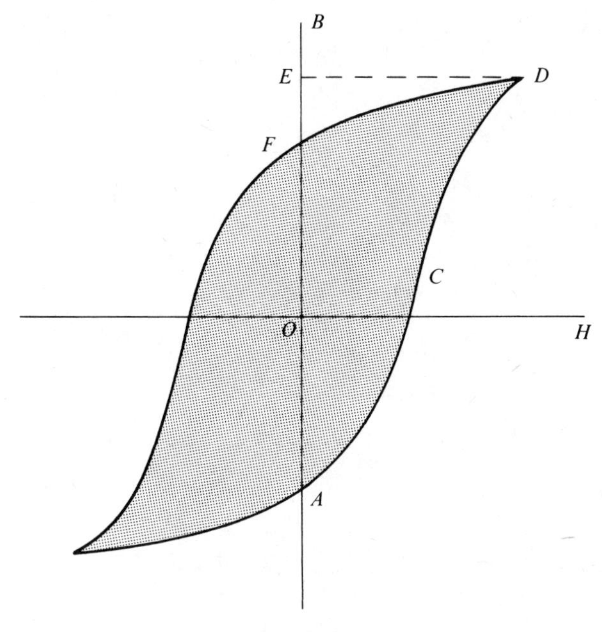


Fig. 7.5

We now suppose the current to be increased by an infinitesimal amount $\mathrm{d}I$, to cause an increase $\mathrm{d}B$ in the flux density, the change taking place in a time interval $\mathrm{d}t$. The resultant change of flux through the winding will be $A\,\mathrm{d}B$ and this will induce in the winding an e.m.f. $AN_1\,\mathrm{d}B/\mathrm{d}t$, in a direction to oppose the flow of $I$. During the time $\mathrm{d}t$ the charge flowing through the winding is $I\mathrm{d}t$, so the work $\mathrm{d}W$ done against the induced e.m.f. is

$$\mathrm{d}W = AN_1 \frac{\mathrm{d}B}{\mathrm{d}t} I\mathrm{d}t = AN_1 I\mathrm{d}B \tag{7.10}$$

Substituting from (7.9), we may write

$$\mathrm{d}W = 2\pi rAH\mathrm{d}B \tag{7.11}$$

However, $2\pi rA$ is the volume of the material, so we have finally

$$\text{Work done per unit volume of material} = H\mathrm{d}B$$

We turn now to the hysteresis loop of fig. 7.5 and consider the magnetization of unit volume of the material from state $A$ to state $D$. Throughout this change both $H$ and $\mathrm{d}B/\mathrm{d}t$ are positive, so the work done on the material is

$$\int H\mathrm{d}B = \text{area enclosed between } ACDE \text{ and the axis of } B$$

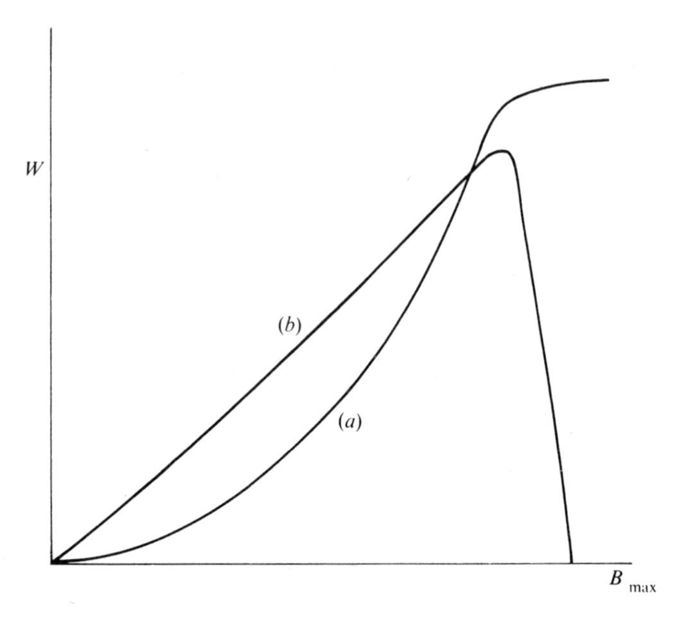


Fig. 7.6

When the material is partially demagnetized from state $D$ to state $F$, $H$ is still positive but $\mathrm{d}B/\mathrm{d}t$ is now negative and work is taken back from the material, of amount equal to the area of $DEF$. The nett result is that, for the complete change represented by the portion of the hysteresis loop lying to the right of the axis of $B$, the work done on the material is equal to the area enclosed by this half of the loop and the $B$-axis. An exactly similar argument applies to the other half of the loop so, for a complete cycle of magnetization, we have

$$\text{Work supplied per unit volume} = \text{area enclosed by the hysteresis loop} \qquad (7.12)$$

This work appears as heat dissipated in the material. It represents a by no means negligible loss of power in transformers and other alternating-current apparatus.

It is clear that the hysteresis loss per cycle will increase as the maximum flux density in the material gets larger and the general relation between the loss $W$ and $B_{max}$ is shown by curve ($a$) of fig. 7.6. Many years ago an empirical law was enunciated by Steinmetz, to the effect that the centre portion of this curve could be represented by the expression

$$W = kB_{max}^{1\cdot 6} \qquad (7.13)$$

However, for many of the magnetic materials now available this relation is not valid and it is better not to rely on it.

So far we have considered only toroidal specimens, but if we think of a piece of material in the form of a short cylinder, there are two different ways in which it may be taken through a cycle of magnetization. One is the method already considered, in which the cylinder is placed in an alternating field directed along its axis. The alternative is to have a constant magnetic field of fixed direction, initially along the axis of the specimen. If the specimen is now turned through 360 degrees about a line perpendicular to its axis, it will undergo a 'directional' cycle of magnetization and the hysteresis loss is found to be quite different in this case. Its variation with $B_{max}$ is shown by curve (*b*) of fig. 7.6. A purely directional cycle rarely occurs in practice. Even in the armature of a rotating machine there is normally a large central hole, so that the flux is carried round an annulus of iron.

### 7.2.6 Classes of magnetic material

The range of magnetic materials available to the electrical engineer is extremely wide and is continually increasing. We shall make no attempt to describe the materials in detail, but shall indicate the properties of a few broad classes.

For most applications, particularly those involving alternating currents, it is desirable that a material should have high permeability and low hysteresis loss – properties which tend to run together. Other useful attributes are high maximum flux densities when saturation occurs and high resistivities which, as we shall see later (§10.3), reduce power loss resulting from eddy currents. Substances possessing these general properties are known as *soft magnetic materials*. Within the range of these materials, the following groups may be distinguished.

(*a*) *Iron, steel and silicon–iron alloys*. Of all magnetic materials, this group finds the most extensive use. The materials are relatively inexpensive and have high permeabilities until saturation occurs and high maximum flux densities. Iron and steel, whether cast or rolled, are used for cores to be magnetized by direct current. For alternating-current power equipment an alloy consisting of essentially pure iron and between one and four per cent of silicon is commonly used. This reduces the hysteresis loss and increases the permeability at low flux densities. It also increases the resistivity. To reduce eddy currents the alloy is rolled into sheets about 0.5 mm thick and the magnetic cores are built up from stampings, which

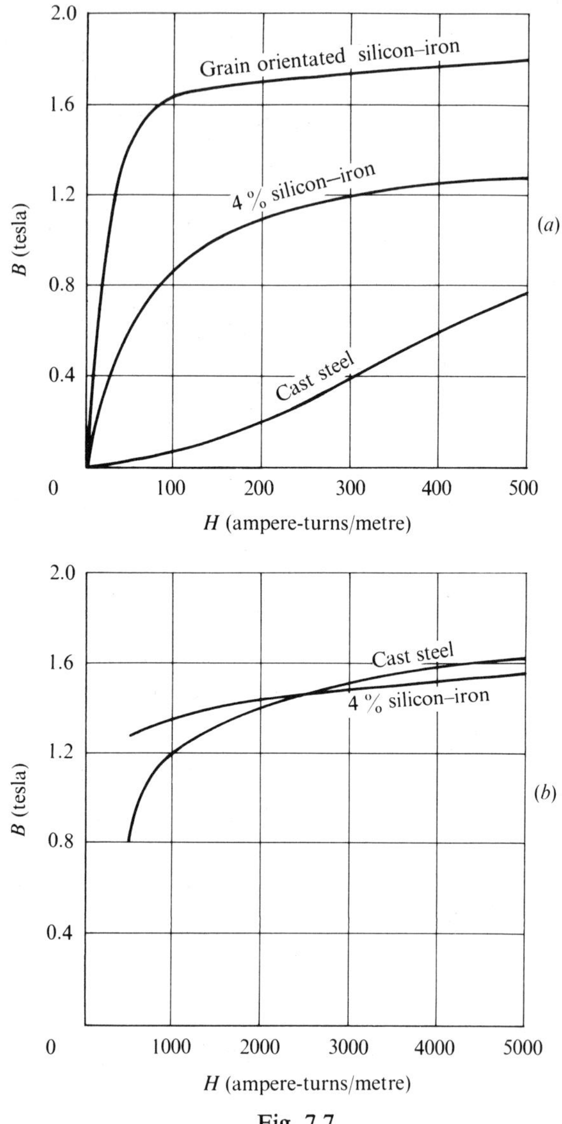


Fig. 7.7

are insulated from each other. Curves for a few of these materials are shown in fig. 7.7.

(*b*) *Nickel–iron alloys.* These alloys, which often contain small quantities of other elements, are much more expensive than the materials of group (*a*) and have much higher permeabilities than the latter, particularly at low

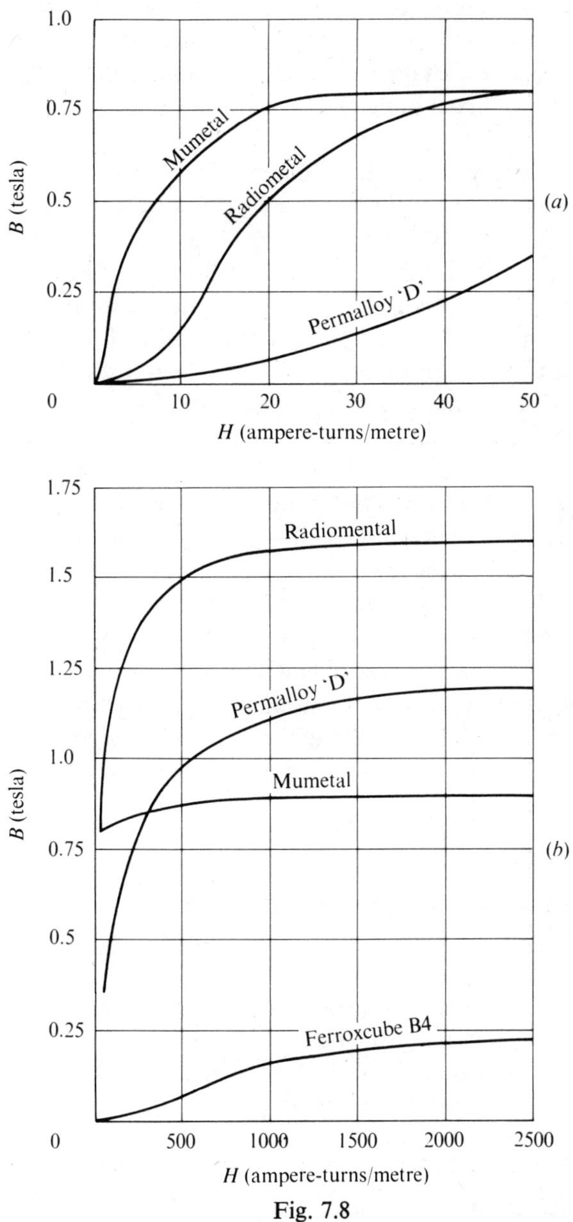
1.0
0.75
0.5
0.25
0
B (tesla)
Mumetal
Radiometal
Permalloy 'D'
(a)
10
20
30
40
50
H (ampere-turns/metre)
1.75
1.5
1.25
1.0
0.75
0.5
0.25
0
B (tesla)
Radiomental
Permalloy 'D'
Mumetal
Ferroxcube B4
(b)
500
1000
1500
2000
2500
H (ampere-turns/metre)

**Fig. 7.8**

flux densities. They are widely used in electronic engineering for magnetic shielding and as laminations for small transformers to operate at frequencies in the audio range. Their maximum flux densities are considerably less than those obtainable with the materials of group (*a*). Curves for a few of these materials are shown in fig. 7.8.

(*c*) *Ferrites.* These are non-metallic materials which are formed into the required shapes by pressing or extrusion, followed by sintering. In general, their magnetic properties are not as good as those of materials in groups (*a*) and (*b*), but their resistivities are about a million times as great. In consequence, eddy-current losses are almost negligible in ferrite cores, which can therefore be used at frequencies up to some tens of megahertz. Many different ferrites are manufactured to cover different uses. Ferroxcube B4 would be suitable for the frequency range 5–10 MHz; its $B$–$H$ curve is shown in fig. 7.8(*b*).

*Hard magnetic materials* are those which have hysteresis loops of large area, giving high retentivity and coercivity. They are used mainly for the manufacture of permanent magnets. For inexpensive magnets, steel containing small percentages of tungsten, chromium and cobalt is frequently used, but very much better magnetic properties can be obtained from alloys of aluminium, nickel, cobalt and copper with, sometimes, one per cent of niobium. Certain ferrites also make very good permanent magnets.

The permeabilities of the hard magnetic materials are very much lower than those of the soft materials. Representative curves are shown in fig. 7.9.

### 7.2.7 The magnetic circuit

Consider the situation shown in fig. 7.10, where a magnetic core in the form of a hollow square is magnetized by current $I$ flowing through a short coil of $N$ turns. We suppose the core to be made of soft magnetic material which, for the time being, we assume to have constant relative permeability $\mu_r$, which is not less than 1000. Let $C$ and $D$ be any two points on the surface of the material, outside the magnetizing coil, and let the magnetic potentials at these points be $U_C$ and $U_D$ respectively. Consider two adjacent paths from $C$ to $D$, which do not thread the coil, with one path lying wholly in the material and the other wholly in air, for which $\mu_r$ is very nearly equal to unity. Then

$$\underset{\text{Iron}}{\int_C^D \boldsymbol{H}\cdot \mathrm{d}\boldsymbol{s}} = \underset{\text{Air}}{\int_C^D \boldsymbol{H}\cdot \mathrm{d}\boldsymbol{s}} = U_C - U_D$$

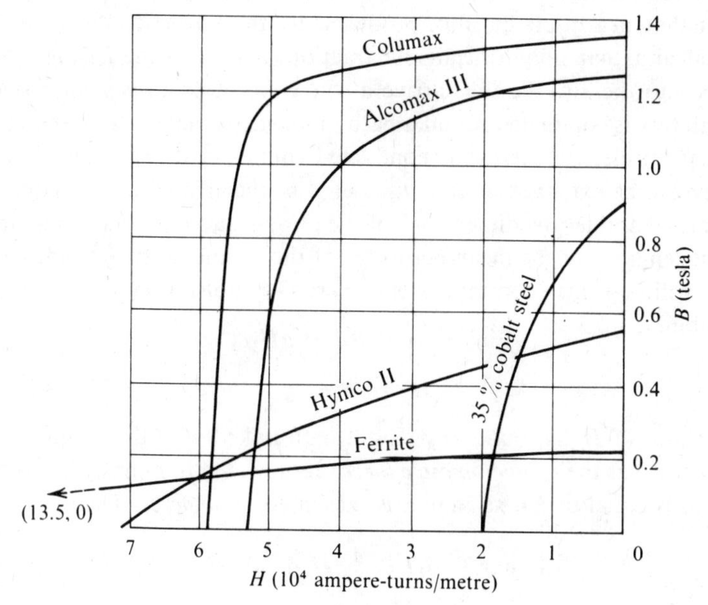


Fig. 7.9

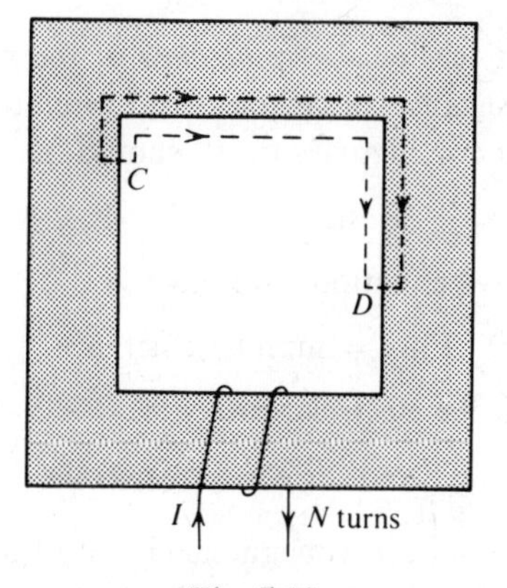


Fig. 7.10

The lengths of the two paths are roughly the same, so the average value of $H$ must be approximately the same in the two cases. Taking account of the high relative permeability of the material, this means that the flux density in the material must be something like 1000 times as great as that in air, for these two paths. Since we can apply this argument to any pair of points on the surface of the material and, by a simple extension, to pairs of paths which thread the coil as well as to pairs which do not, we reach the conclusion that nearly all the flux produced by the coil is confined to the material and that only a negligible fraction, known as the *leakage flux*, escapes into the air. We thus arrive at the concept of a *magnetic circuit*, in which flux is conducted round a path in magnetic materials, in the same way that an electric current is conducted roung a path in a metal. The analogy can be extended as follows. Let $A$ be the area of cross-section of the material and let the dimensions of the cross-section be small compared with the length $L$ of the mean perimeter of the circuit. Then $H$, and therefore $B$, will be nearly constant over the cross-section. If $\mu$ $(= \mu_0\mu_r)$ is the permeability,

$$\text{Flux} = \phi = BA = \mu HA$$

and

$$H = \phi/\mu A \tag{7.14}$$

The integral of $\boldsymbol{H}\cdot\mathrm{d}\boldsymbol{s}$ round any closed path linked with the magnetizing coil is known as the *magnetomotive force* (abbreviated to m.m.f.) round the path and is equal to $NI$, so m.m.f. is measured in *amperes*. Then

$$\text{m.m.f.} = NI = \oint_s \boldsymbol{H}\cdot\mathrm{d}\boldsymbol{s} = HL$$

and, substituting from (7.14),

$$\text{m.m.f.} = \phi(L/\mu A) \tag{7.15}$$

The quantity $L/\mu A$ is termed the *reluctance* of the magnetic circuit; it is analogous to the resistance of an electrical circuit for which, with the same dimensions and conductivity $\sigma$, we should have

$$\text{Resistance} = L/\sigma A \tag{7.16}$$

Finally, from the above equations, we have

$$\text{Flux} = \text{m.m.f./reluctance} \tag{7.17}$$

which corresponds to the electrical case

$$\text{Current} = \text{e.m.f./resistance}$$

These equations are valid for a magnetic circuit of any shape, so long as the permeability can be taken to be constant, the area of cross-section is constant and the dimensions of the cross-section are small compared with the length round the circuit. If the last two conditions are not satisfied we can still, in principle, use the methods of chapter 6 to determine how the flux distributes itself round the circuit, and thus to find the value of the flux density at any point. However, this is rarely worth while in view of the assumption of constant permeability. We shall see later how this assumption can be avoided.

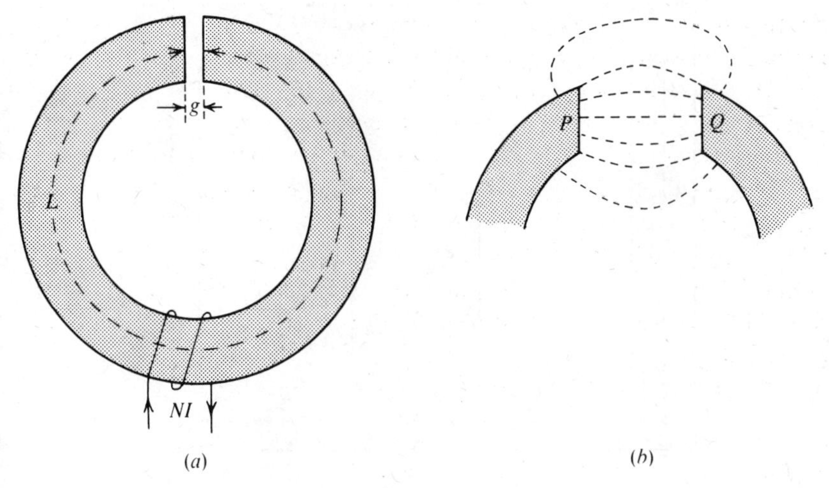


Fig. 7.11

### 7.2.8 Reluctances in series or parallel

As an example of reluctances in series, we consider a ring of magnetic material of cross-sectional area $A$, of mean perimeter $L$ and with a small air gap of width $g$ as in fig. 7.11($a$). For the material the permeability is $\mu_r\mu_0$, while for the air it is $\mu_0$. The m.m.f. is $NI$ ampere-turns. Since $g$ is small we assume that the flux $\phi$ passes across the gap without appreciable fringing. Then $B$ is equal to $\phi/A$ throughout the circuit. Within the material

$$H_{\text{mat.}} = B/\mu_0\mu_r = \phi/A\mu_0\mu_r$$

but, in the air gap, $\quad H_{\text{air}} = B/\mu_0 = \phi/A\mu_0$

For the complete circuit

$$\text{m.m.f.} = NI = (\phi L/\mu_0\mu_r A) + (\phi g/\mu_0 A)$$

or

$$\phi = \frac{NI}{(L/\mu_0\mu_r A) + (g/\mu_0 A)} \tag{7.18}$$

This equation shows that, as we might expect from the electrical analogy, reluctances in series must be added to each other. It also shows the very large effect exerted by the small air gap. If $\mu_r$ is 1000, an air gap of one millimetre is equivalent to a path length of one metre in the material.

If the air gap is relatively large, the assumption that there is no fringing of the flux is no longer valid and we may expect the distribution of flux at the gap to be roughly as shown in fig. 7.11($b$). However, in this case, the reluctance of the magnetic material will form a quite negligible proportion of the total reluctance and we may consider the portions of the material

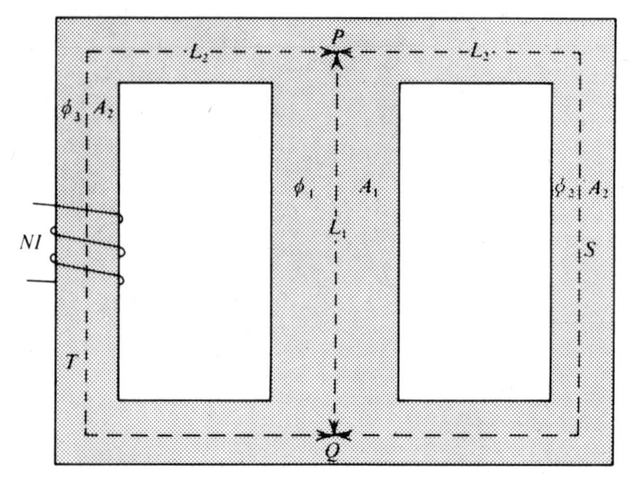


Fig. 7.12

in the vicinity of the gap, $P$ and $Q$ in the figure, to be magnetic equipotential surfaces. The magnetic potential difference between them will be $NI$ and the flux passing from one to the other can be found by the methods of chapter 6, using a computer if analytical methods are inapplicable.

To illustrate the combination of reluctances in parallel we take the arrangement shown in fig. 7.12, where a transformer core has a magnetizing winding of $NI$ ampere-turns on one of its outer limbs. We shall ignore the effects caused by flux turning round corners, and take mean path lengths $L_1$ and $L_2$ as shown. The permeability throughout the material is $\mu$. The areas of cross-section are $A_1$ for the central branch and $A_2$ for the remainder of the core. Thus the reluctances are

$$R_1 = L_1/\mu A_1$$

for the central branch and $$R_2 = L_2/\mu A_2$$

for each of the outer branches. Let $U_P$ and $U_Q$ be the magnetic potentials of points $P$ and $Q$ respectively, $\phi_1$ the flux in the central branch, $\phi_2$ the flux in the branch $PSQ$ and $\phi_3$ that in the branch $PTQ$. Since the source of all the flux is the winding on $PTQ$,

$$\phi_3 = \phi_1 + \phi_2$$

$$\phi_1 = (U_P - U_Q)/R_1, \quad \phi_2 = (U_P - U_Q)/R_2$$

and $$\phi_3 = (U_P - U_Q)\left(\frac{1}{R_1} + \frac{1}{R_2}\right)$$

Thus the two reluctances $R_1$ and $R_2$, which are in parallel, combine by the

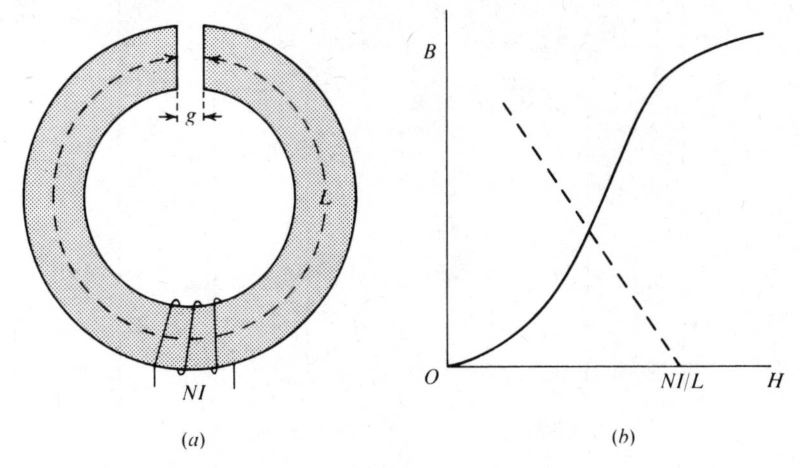


Fig. 7.13

same rule as resistances in parallel, and their total reluctance is

$$R_1 R_2/(R_1+R_2)$$

For the complete system, we then have

$$\phi_3\{R_2+(R_1 R_2)/(R_1+R_2)\} = NI$$

also $$\phi_1 = \phi_3 R_2/(R_1+R_2) \quad \text{and} \quad \phi_2 = \phi_3 R_1/(R_1+R_2)$$

so the fluxes in all branches are known.

### 7.2.9 Calculations taking account of the variation of $\mu$

Although the method outlined in the preceding section is often useful for rough calculations, its accuracy is limited by the assumption of a permeability which is independent of flux density. We now re-examine the same two problems, taking account of the fact that $B$ is not proportional to $H$. Our starting point must, of course, be a graph showing the relation between $B$ and $H$.

In fig. 7.13(*a*) we return to the ring of magnetic material with a small air gap, magnetized by a coil with $NI$ ampere-turns. The path length in the material is $L$ and the width $g$ of the air gap is small enough for fringing to be neglected. Then the flux density $B$ is the same in the air as in the iron and the value of $H$ in the air is $B/\mu_0$. Hence, for a complete circuit of the ring, if $H_{\text{mat.}}$ is the value of $H$ in the material,

$$\oint_s \boldsymbol{H}\cdot \mathrm{d}\boldsymbol{s} = (Bg/\mu_0)+H_{\text{mat}}\, L = NI \tag{7.19}$$

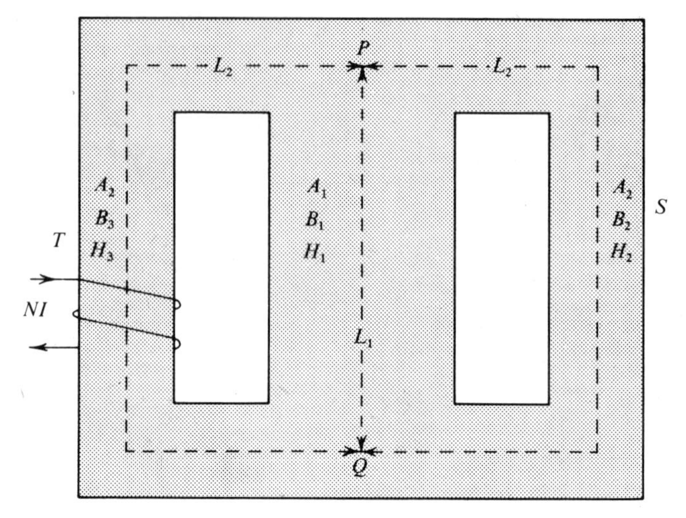


Fig. 7.14

We shall also have the relation between $H$ and $B$ in the form of a curve such as fig. 7.13($b$). If our problem is to find $NI$ in order to produce a given $B$, we can determine $H$ from the curve and (7.19) then enables us to calculate $NI$. On the other hand, we may be given $NI$ and asked to find $B$. In that case the straight line represented by (7.19) is plotted on the $B$–$H$ curve for the material, and its intersection with the curve gives the required value of $B$.

The second problem of the preceding section is reproduced in fig. 7.14, where the areas of cross-section and the values of $B$ and $H$ for each branch are indicated. Suppose first that we are given the flux density $B_2$ in branch $PSQ$ and are asked to find $NI$. For this branch we can immediately find $H_2$ from the given curve, such as fig. 7.13($b$). If $U_P$ and $U_Q$ are the magnetic potentials at $P$ and $Q$ respectively

$$U_P - U_Q = H_2 L_2$$

and

$$H_1 = (U_P - U_Q)/L_1 = H_2 L_2 / L_1$$

Using this value of $H_1$, we find $B_1$ from the curve.

The fluxes in $PQ$ and $PSQ$ are $B_1 A_1$ and $B_2 A_2$ respectively, so the total flux in $QTP$ is $(B_1 A_1 + B_2 A_2)$ and the flux density $B_3$ in this branch is $(B_1 A_1 + B_2 A_2)/A_2$. Again turning to the curve, we can find $H_3$. Finally, for a complete circuit such as $PQTP$, we have

$$H_3 L_2 + H_1 L_1 = NI$$

and the problem is solved.

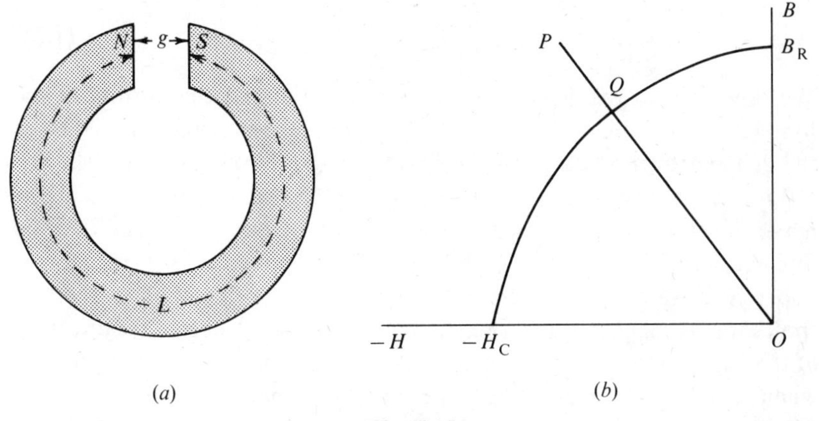


Fig. 7.15

If, on the other hand, we are given $NI$ and are asked to find $B$ in one of the branches, we cannot proceed by the foregoing method, since we have no starting point in the value of $B$ or $H$ in any branch. We must then resort to a trial-and-error method, assuming a value for $B$ in one branch and working back to $NI$. The method of the preceding section, using a constant average value for $\mu$, will provide an approximate first choice for $B$.

### 7.2.10 Permanent magnets

If a toroidal specimen of a material which exhibits hysteresis is magnetized to saturation, and the magnetizing current is reduced to zero, the flux density will fall to some value $B_R$, which we have termed the retentivity, and will retain this value indefinitely. The material has become permanently magnetized. However, a magnetized toroid is of little practical value, since we cannot make use of the flux which it contains. A useful permanent magnet must have an air gap within which the flux can be used. We therefore consider the magnet shown in fig. 7.15(*a*), where $L$ is the length of path in the material and $g$ the width of the air gap. For many applications $g$ will not be sufficiently small to prevent fringing of the flux but, for the time being, we ignore this complication and take $g$ to represent the effective gap width. Then, after the magnetizing current has been switched off, we have

$$B_{\text{air}} = \mu_0 H_{\text{air}} = B_{\text{mat.}} \tag{7.20}$$

and

$$\oint_s \boldsymbol{H} \cdot \mathrm{d}\boldsymbol{s} = gH_{\text{air}} + LH_{\text{mat.}} = 0 \tag{7.21}$$

Eliminating $H_{\text{air}}$ from these equations,

$$\frac{B_{\text{mat.}}}{H_{\text{mat.}}} = -\frac{\mu_0 L}{g} \tag{7.22}$$

We now turn to the *B–H* curve for the material and the portion of the hysteresis loop which is of interest, that lying in the second quandrant, is the demagnetization curve shown in fig. 7.15(*b*). Plotting the straight line *OP* represented by (7.22) on this curve, we see that the final state of the material will be that indicated by the point of intersection *Q*, and that the flux density will be appreciably less than $B_R$. The reason for this becomes apparent if we take $U_N$ and $U_S$ to be the magnetic potentials at the two 'poles' of the magnet *N* and *S* respectively. It is then clear that the direction of the field in the air gap is opposite to that in the material so that, for the whole circuit, the field in the material is less than it would have been without the gap. This is sometimes expressed by saying that the poles of the magnet exert a demagnetizing effect on the material but, from the standpoint adopted in this book, no quantitative significance is to be attached to the poles.

In the foregoing discussion it has been assumed that the magnet is made of a single material but, except for the smaller magnets, this is not the usual practice. Hard magnetic materials are expensive and are difficult to machine. It is therefore economical to use the smallest possible quantity of magnet material, in a simple geometrical form, and to lead the flux to the air gap through soft-iron pole pieces which can be machined without difficulty to any desired shape. An example of a magnet of this kind is indicated in fig. 7.16(*a*) and we need some criterion to help us to choose the best proportions for the soft iron and the magnet material.

As a first step we shall assume that there is no leakage or fringing of flux and that the permeability of the soft iron is so large that the reluctance of the iron can be neglected. Let $g$ be the width and $A_g$ the area of cross-section of the gap, $B_g$ the required flux density and $H_g = B_g/\mu_0$ the corresponding field strength. All of these values will be stipulated. Let $L$, $A$, $B$ and $H$ be the corresponding values for the magnet material, which we wish to find. We have the following equations for the magnetized material

$$\text{Total flux} = B_g A_g = BA \tag{7.23}$$

$$\oint_s \boldsymbol{H}\cdot d\boldsymbol{s} = gH_g + LH = 0 \tag{7.24}$$

$$\text{Volume of magnetic material} = LA \tag{7.25}$$

Substituting for $L$ and $A$ in (7.25)

$$\text{Volume of material} = \frac{H_g g B_g A_g}{(-H)B} \tag{7.26}$$

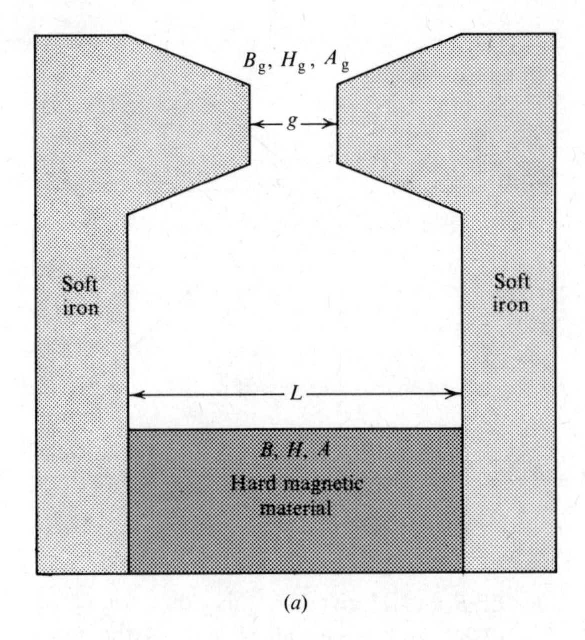


(*a*)

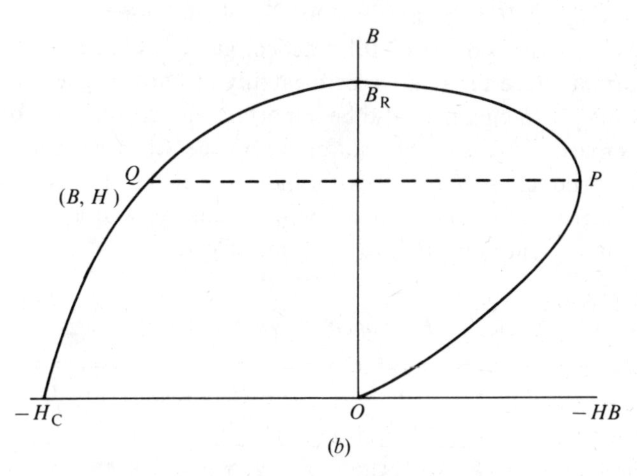


(*b*)

Fig. 7.16

Thus, if the volume of material is to be as small as possible, $(-H)B$ should be a maximum.

Turning now to fig. 7.16(*b*), the curve on the left of the axis of $B$ is the ordinary demagnetization curve for the material. On the right of the axis $(-H)B$ is plotted as a function of $B$, and has a maximum at the point $P$, corresponding to the state represented on the demagnetization curve by $Q$.

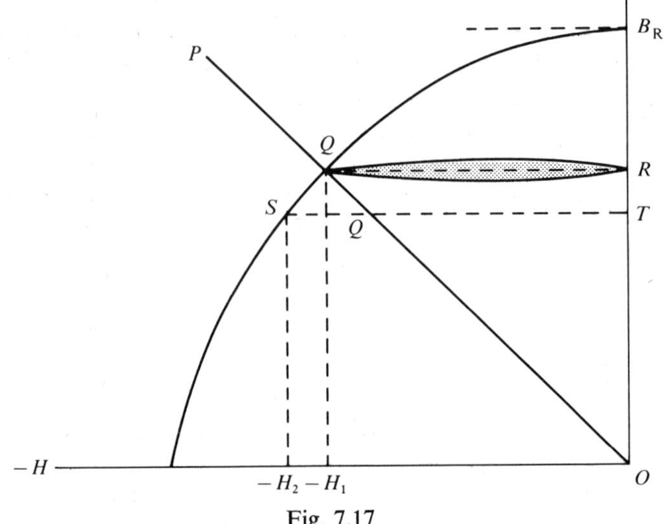


Fig. 7.17

Using the values of $B$ and $H$ given by this point, we can find $A$ and $L$ from (7.23) and (7.24) and these values give us the most economical magnet. Setting $(-H)B$ to a maximum is often known as *Evershed's criterion*. While it provides a useful practical guide, leakage flux is often quite important, since the relative permeability of hard magnet materials is not very high. Fringing may also be important and corrections for these two factors must be made in the light of experience. Clearly the maximum value of the product $(-H)B$ is a useful indicator of the effectiveness of a magnetic material and figures for a few materials, as well as their coercivities $H_C$ and retentivities $B_R$ are given in table 7.1.

Table 7.1 *Permanent magnet materials*

| | Composition | $H_C$ A m$^{-1}$ | $B_R$ tesla | $(-HB)_{max.}$ J m$^{-3}$ |
|---|---|---|---|---|
| Cobalt steel | 35 % Co, 3.5 % Cr, 3 % W, 1 % C | $2 \times 10^4$ | 0.90 | 7.6 |
| Alnico V | 13.5 % Ni, 8 % Al, 24 % Co, 3 % Cu | $5.2 \times 10^4$ | 1.27 | 44 |
| Columax | 13.5 % Ni, 8 % Al, 24 % Co, 3 % Cu. Crystals aligned | $6.0 \times 10^4$ | 1.35 | 62 |
| Magnadur II | Anisotropic $BaFe_{12}O_{19}$ | $13.5 \times 10^4$ | 0.38 | 24 |

Corresponding to the minor hysteresis loops discussed earlier in connection with incremental permeability, there are similar loops on the demagnetization curves of hard magnetic materials. Referring to fig. 7.17, suppose a magnet has been magnetized to saturation and the magnetizing current has then been switched off. The state of the material will be represented by point $Q$, where the line $OP$ is the plot of an equation like (7.22). The effective demagnetizing field in the material is now $-H_1$. Next let the current in the magnetizing winding be increased in a direction to reduce the total demagnetizing field in the material to zero. This causes the state point to traverse the lower branch of the curve from $Q$ to $R$. Finally, when the current is reduced to zero once more, the state point moves back to $Q$ along the upper branch of the curve. Minor loops such as the one between $Q$ and $R$ are often termed *recoil loops*. They are very narrow and, for many purposes, can be approximated by straight lines which are roughly parallel to the tangent to the main curve at $B_R$.

There are many applications of permanent magnets in which the constancy of the flux density is more important than its magnitude. This is the case in moving-coil meters of all kinds and, to ensure constancy, we can make use of a recoil loop in the following way. If, when the magnet is in state $Q$, it experiences a further demagnetizing field (perhaps as a result of accidental proximity to another magnet), there will be a serious loss of flux density, since $Q$ is on a steep part of the curve. Moreover, the initial flux will not be regained when the disturbing field is removed, since the state point will move along a recoil loop. To overcome this difficulty the demagnetizing field is deliberately increased to some value $-H_2$, which is slightly greater than any field likely to be encountered by accident. When the current is switched off, the state point travels along a recoil loop, which we approximate by the straight line $ST$, to reach the final state $Q'$ at the intersection of $ST$ and $OP$. This is the state in which the magnet is used. Any accidental additional field, in either direction, will cause the state point to move along $ST$ and, since this line is nearly parallel to the axis of $H$, the resulting changes in flux density will be small. Moreover, the change is reversible and the state point will move back to $Q'$ when the disturbing field is removed. Thus, at a sacrifice of flux density which need not be great, the performance of the magnet has been stabilized.

A further consequence of the form of the recoil loops concerns the stage of manufacture at which a magnet assembly should be magnetized. Consider for example, the case of a rotating machine with a permanent field-magnet. The final air gap in the magnetic circuit will be very much smaller than the gap which exists before the armature is inserted. If, therefore, the assembly is magnetized after the armature is in position, the flux density will be considerably greater than would be the case if the magnet were first magnetized and the armature inserted subsequently. In general, it is advantageous to magnetize when an assembly is complete.

## 7.3 Dielectric materials

### 7.3.1 Classification of dielectric materials

The great majority of dielectric materials are linear in the sense that the relative permittivity is independent of the applied electric field strength. The relative permittivity usually lies between one and ten and may vary quite widely with frequency. These materials generally exhibit rather small hysteresis, which may nevertheless cause serious loss of power at frequencies in excess of 1 MHz.

A small number of substances, such as Rochelle salt, triglycine sulphate and barium titanate, have relative permittivities which may exceed $10^4$ and are known as *ferroelectric materials*. Their relative permittivities often depend very strongly on temperature and, when the materials are crystalline, may be quite different in different crystal directions. When the electric displacement $D$ is plotted against the electric field strength $E$, ferroelectrics exhibit hysteresis loops similar to those found in ferromagnetic $B$–$H$ curves. Many ferroelectrics have associated piezoelectric properties and are of technological importance on this account. The principal use of ferroelectrics for their dielectric properties is in the manufacture of capacitors of very small size.

### 7.3.2 The measurement of dielectric properties

The permittivity of a material is, by definition, the ratio of $\boldsymbol{D}$ to $\boldsymbol{E}$ in that material. While it is a simple matter to apply to the material an electric field of known strength, there is no simple way in which $\boldsymbol{D}$ can be measured directly, so the permittivity is usually deduced from a measurement of capacitance.

We have previously seen (§3.7.4) that the capacitance per unit area between two infinite parallel plane conductors in free space, separated by a distance $d$, is $\epsilon_0/d$. If, however, the space between the plates is completely filled with material of relative permittivity $\epsilon_r$, the ratio of $\boldsymbol{D}$ to $\boldsymbol{E}$ in this region will be increased by this factor. We may state this in another way by saying that, for a given potential differences between the plates, the charges on them have been increased by a factor $\epsilon_r$ and that, in consequence, the capacitance per unit area is now $\epsilon_0\epsilon_r/d$. If, therefore, this capacitance can be measured, $\epsilon_r$ can be found.

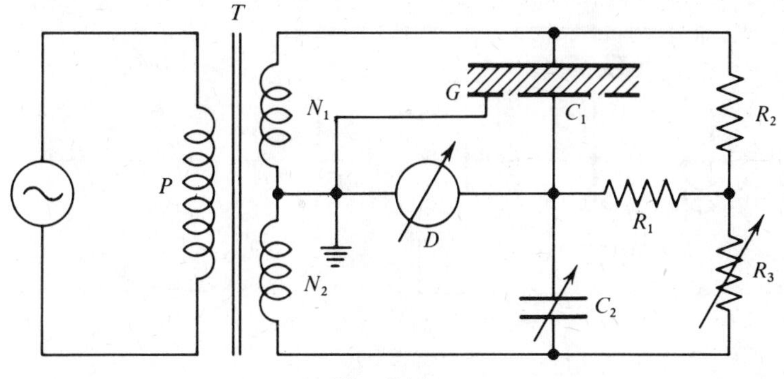


Fig. 7.18

In practice the plates must be of finite size, but edge effects can be avoided by the use of circular plates with guard rings (§3.7.8). It is important to avoid any air space between the material and the plates, and this may conveniently be done by ensuring that the surfaces of the material are smooth, by using tinfoil for the conducting plates and by attaching the tinfoil to the material with a very thin film of vaseline. A convenient size for the sample of material might be a disc about 10 cm in diameter and a few millimetres thick. If $\epsilon_r$ is less than ten, the capacitance will then be a few tens of picofarads, so care must be taken to avoid errors resulting from stray capacitances.

It is not our intention to discuss in any detail the problems associated with the measurement of small capacitances, but we shall have occasion to refer later to an alternating current bridge which can be used for this purpose and it is convenient to give a brief description of it here. The circuit is shown in fig. 7.18, where $T$ is a transformer on which there are three windings. The primary widing $P$ is connected to an oscillator, which supplies power to operate the bridge. The other two windings, with numbers of turns $N_1$ and $N_2$ respectively, form the ratio-arms of the bridge and every effort is made to secure the closest possible coupling between these two windings. It can be shown that, if the couplings were perfect, the bridge would be completely immune from the effects of stray capacitance. $D$ is a detector, which may conveniently be an oscilloscope, and $C_2$ is a calibrated variable air-capacitor. $C_1$ is the capacitor formed by the sample and it will be observed that, when the bridge is balanced, the guard ring $G$ is at the same potential as the electrode which it surrounds. The network formed by $R_1$, $R_2$ and $R_3$, which need not concern us here, is provided to balance any imperfections resulting from power losses in the capacitors. $R_1$ is usually of the order of 100 k$\Omega$, while $R_2$ and $R_3$ might each be a few

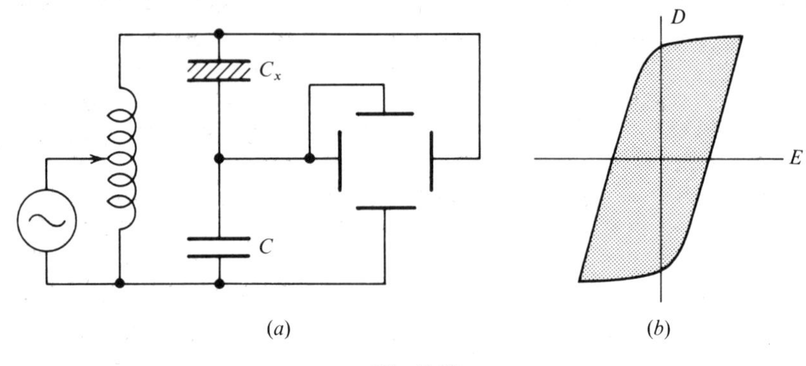


Fig. 7.19

kilohms. When balance has been obtained by adjustment of $C_2$ and $R_3$, it can be shown that

$$C_1/C_2 = N_2/N_1 \tag{7.27}$$

Since $\epsilon_r$ varies with frequency, measurements of $C_1$ must be made over a range of frequencies. If the core of the transformer $T$ is made of thin nickel–iron stampings, the bridge can be used from about 10 Hz to 10 kHz. With very simple apparatus a balance sensitivity of 0.01 pF can readily be achieved and much greater precision is possible when suitable precautions are taken. For measurements at frequencies above 10 kHz transformers with ferrite cores may be used but, for the highest frequencies, other methods are needed and we shall not consider these.

From what has been said about the non-linearity of ferroelectrics, it will be clear that bridge methods would not be suitable for the measurement of their properties. The hysteresis loop for a ferroelectric can be traced on an oscilloscope using the circuit of fig. 7.19(*a*), while a typical loop for barium titanate ($BaTiO_3$) is shown in fig. 7.19(*b*).

### 7.3.3 Electrets

Many dielectric materials contain molecules which have permanent dipole moments and when such materials are placed in a strong electric field there will be a tendency for the dipoles to become aligned in the direction of the field. However, internal molecular forces oppose the alignment which, for solids at room temperature, is usually far from complete. If the temperature of the material is raised while it is in the field, the opposing forces decrease and much more perfect alignment can then be secured. Finally, if the material is cooled to room temperature, while still in the field, the more perfect alignment persists and does not disappear even when the external field is removed. We then have the electrostatic analogue of a permanent magnet; it is termed an *electret*.

The behaviour of electrets is even more complicated than that of permanent magnets. The magnetism of the latter can result only from magnetic dipoles (or amperian currents) since separate magnetic poles do not exist. In the electrostatic case, however, separate positively and negatively charged particles do exist, are present in the dielectric material and can be externally transferred to the surface of the material. It is outside the scope of this book to describe the different processes by which electrets may be produced, but it may be said that the 'freezing-in' of aligned dipoles, referred to in the preceding paragraph, does not by itself lead to electrets which maintain their strength for long periods of time. On the other hand, processes which cause separate positive and negative charges to become trapped on opposite surfaces of a dielectric sheet can produce electrets which will provide strong, stable electrostatic fields for many years. Electrets of this kind are finding increasing application in industry: they are used, for example, in the construction of electrostatic microphones.

### 7.3.4 Anisotropic materials

Apart from one brief reference to the fact that some ferroelectric crystals have different properties in different directions, it has been tacitly assumed throughout this book that the materials with which we have been concerned were isotropic: that their properties were independent of direction in the material. We should not expect this to be true of a perfect crystal unless it happened to have a simple cubic structure and, in fact, most ordinary materials are isotropic on a macroscopic scale only because they are amorphous and are made up of innumerable crystalites, whose axes are orientated randomly. However, for technological purposes, more and more materials are being produced in the form of large single crystals and many of these have different properties in different directions; they are said to be *anisotropic*. Another anisotropic material which is widely used in the electrical industry is grain-orientated silicon–iron. This is not made as a single crystal, but sheets are rolled in such a way that individual crystallites have their axes aligned in a particular direction, to give favourable magnetic properties in that direction.

Suppose that, in an anisotropic material, we find that the permeability has a maximum value $\mu_1$ in one direction and a minimum value $\mu_2$ in some other direction. If a magnetic field $\boldsymbol{H}$ acts on the material in an arbitrary direction, it will have components along the two directions corresponding

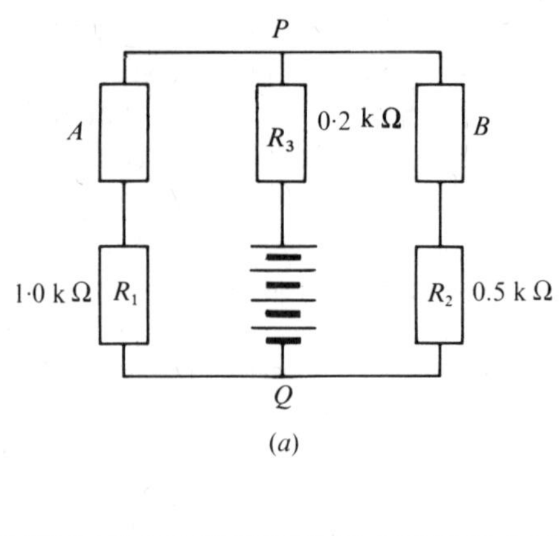


(a)

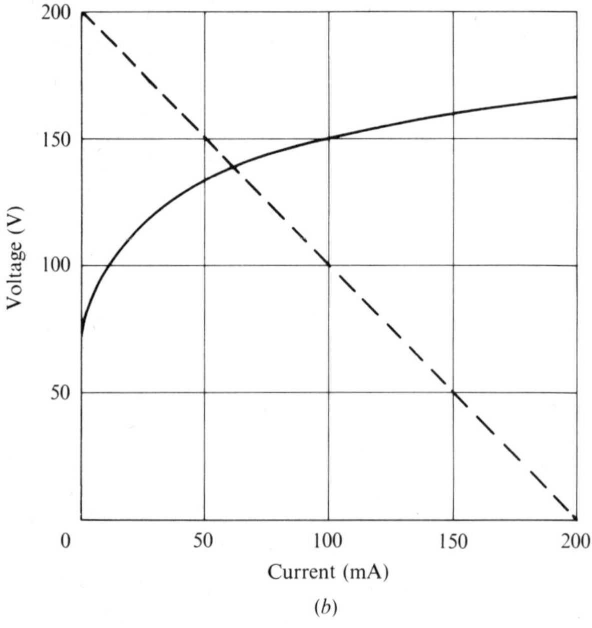


(b)

Fig. 7.20

to $\mu_1$ and $\mu_2$ respectively. However, because the two permeabilities are different, the components of $\boldsymbol{B}$ in the two directions will not be in the same ratio as the components of $\boldsymbol{H}$. It follows from this simple argument that, in general, we must expect the direction of $\boldsymbol{B}$ in an anisotropic material to be different from the direction of $\boldsymbol{H}$. Similarly, we must not expect the directions of $\boldsymbol{D}$ and $\boldsymbol{E}$ to coincide. Further discussion of the complications arising from these facts is outside the scope of this book.

## 7.4 Worked example

In the circuit of fig. 7.20(*a*), *A* and *B* are identical voltage-dependent resistors, for which the relation between voltage and current is shown in fig. 7.20(*b*). Determine the voltage of the battery, if the current through *B* is 100 mA.

*Solution.* From the curve, the voltage across *B* is 150 V. Also the voltage across $R_2$ is 50 V, so the total voltage across each arm is 200 V. If $I$ is the current in the left-hand branch, the voltage across *A* is $200–1000I$ and this is plotted as the broken line in fig. 7.20(*b*). From its intersection with the characteristic curve of *A*, we find that $I = 61$ mA.

The total current flowing through the centre branch is thus 161 mA, and the voltage drop across $R_3$ is 32.2 V. The total battery voltage is therefore 232.2 V.

## 7.5 Problems

1. The magnetic circuit shown in fig. 7.21 is symmetrical and is composed of material having the following characteristics:

| $H$ (A-t m$^{-1}$) | 100 | 200 | 280 | 400 | 600 | 1000 | 1500 | 2500 |
|---|---|---|---|---|---|---|---|---|
| $B$ (T) | 0.51 | 0.98 | 1.20 | 1.37 | 1.51 | 1.65 | 1.73 | 1.78 |

Determine the flux density in the centre limb, and the necessary ampere-turns for a winding on the centre limb,

(*a*) for a flux density of 1.2 T in each 1 mm air gap,

(*b*) for a flux density of 1.2 T in one of the 1 mm air gaps when the other gap is closed.

Neglect flux leakage and fringing effects. (University of London, 1974.)

2. A winding of 500 turns, on the centre limb of the core shown in fig. 7.21, carries a current of one ampere. If the permeability of the material ($\mu_0\mu_r$) may be assumed to have the constant value $4 \times 10^{-3}$ H m, determine the total reluctance of the circuit through which the winding causes flux to flow,

(*a*) when both air gaps are open,

(*b*) when the left-hand air gap is closed.

Hence determine the flux density in the right-hand air gap in each case.

3. A single-loop magnetic circuit, of constant cross-section, has an air gap of length 1 mm. The remainder of the circuit has a mean length of one metre and carries a winding of 500 turns. It is made of material with the following characteristics:

| $H$ (A-t m) | 100 | 200 | 300 | 400 | 500 | 600 | 700 | 800 |
|---|---|---|---|---|---|---|---|---|
| $B$ (T) | 0.51 | 0.98 | 1.24 | 1.37 | 1.45 | 1.51 | 1.55 | 1.59 |

Determine the flux density in the air gap when the current in the winding is 3 A. Neglect fringing and leakage of flux.

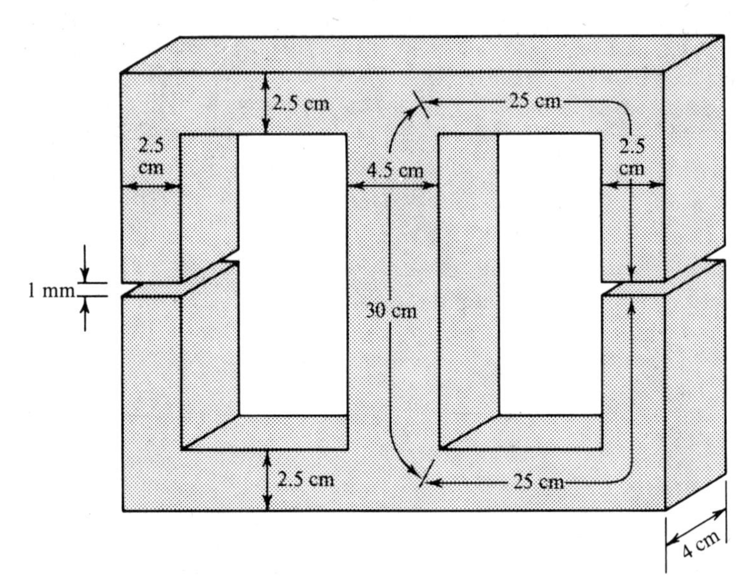


Fig. 7.21

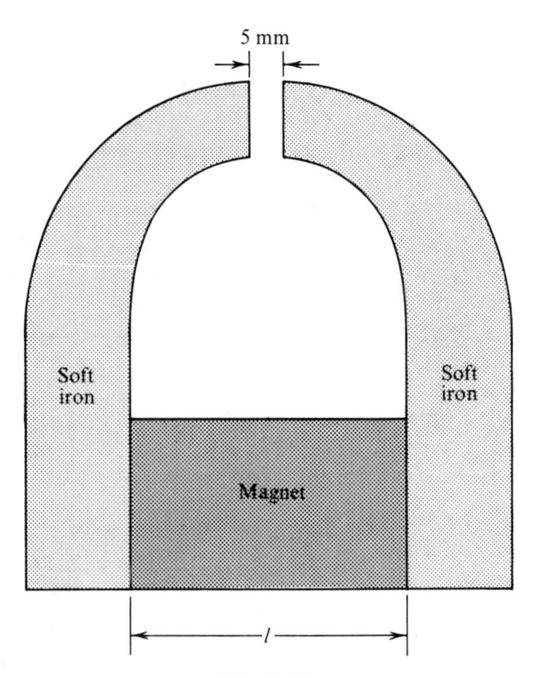


Fig. 7.22

4. In the permanent magnet illustrated in fig. 7.22, the soft-iron pole pieces may be assumed to have negligible reluctance, and fringing and leakage of flux may be neglected. The cross-section of the magnet is 10 $cm^2$ and that of the air gap 5 $cm^2$. The material of the magnet has the following characteristics:

| $H$ (A-t m) | 0 | −2 | −3 | −3.5 | −4 | −4.5 | −5 | −6 | $-6.3 \times 10^4$ |
|---|---|---|---|---|---|---|---|---|---|
| $B$ (T) | 1.11 | 1.09 | 1.08 | 1.04 | 0.98 | 0.88 | 0.74 | 0.31 | 0 |

Estimate $l$ and the resulting flux density in the air gap when the most economical use of the magnet material is achieved.

# 8

# Energy and forces in electric and magnetic fields

## 8.1 Energy storage

### 8.1.1 The energy stored in a charged capacitor

If a capacitor has capacitance $C$ and charges $+Q$ and $-Q$ respectively on its two electrodes, the potential difference between these electrodes is

$$V = Q/C \tag{8.1}$$

The work done in transferring additional charge $\mathrm{d}Q$ from the negative to the positive electrode is

$$\mathrm{d}W = V\mathrm{d}Q = Q\mathrm{d}Q/C \tag{8.2}$$

Hence, the total work done in charging a capacitor is

$$W = \int_0^Q \frac{Q}{C}\,\mathrm{d}Q = \tfrac{1}{2}Q^2/C = \tfrac{1}{2}QV = \tfrac{1}{2}CV^2 \tag{8.3}$$

This work represents energy stored in the capacitor and it can be recovered when the capacitor is discharged.

### 8.1.2 An alternative expression for the stored energy

As a first step in obtaining an alternative expression for the stored energy, we consider a parallel-plate capacitor with plates of infinite area, separated by distance $d$ in free space. As we have seen (§3.7.4), in this capacitor the field between the plates is uniform and, if $V$ is the potential difference between the plates, the field strength $E$ is given by

$$E = V/d \tag{8.4}$$

Again, if $+Q$ and $-Q$ respectively are the charges per unit area of the plates, the displacement $D$ has magnitude

$$D = Q \tag{8.5}$$

Thus, from (8.3) the stored energy $W$ per unit area of the system is

$$W = \tfrac{1}{2}QV = \tfrac{1}{2}DEd \tag{8.6}$$

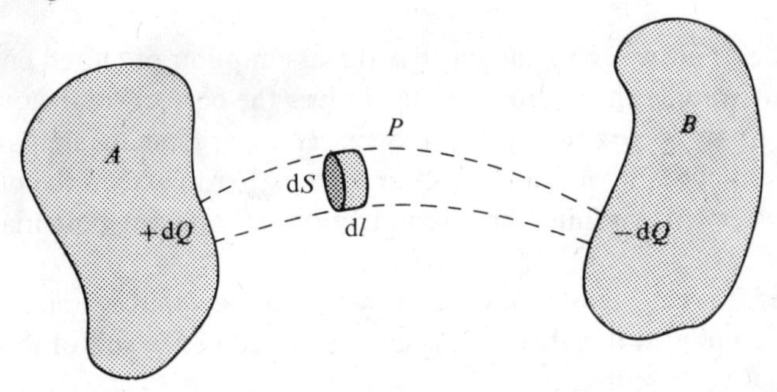


Fig. 8.1

However, in this capacitor, the whole of the electrostatic field lies between the plates and the volume of space in this region, per unit area of the system, is $d$. It thus transpires that if we *assume* energy to be stored throughout the field at a density of $\frac{1}{2}DE$ per unit volume, the total storage calculated on the basis of this assumption will be equal to the energy which, from (8.3), we know actually to be stored in the capacitor.

It is particularly to be noted that we have not asserted that the energy is stored throughout the field and, indeed, when the field is in free space it is difficult to see what meaning could be attached to such an assertion. What we have done is to show that, for a parallel-plate capacitor, there is an alternative method of calculating the stored energy. Clearly, if the new method is to be of much value, we must show that it can be used with electrodes of any shape. This we shall now do.

In fig. 8.1 let $A$ and $B$ be any two conductors in free space and let $V$ be the potential difference between them. Let the dotted lines represent a tube bounded by lines of $\boldsymbol{D}$. Then the same flux of $\boldsymbol{D}$ passes through all cross-sections of this tube and the flux is equal in magnitude to the charges, $+\mathrm{d}Q$ and $-\mathrm{d}Q$ respectively, on which the tube ends. Thus the value of $D$ at any point is equal to $\mathrm{d}Q$ divided by the area of cross-section of the tube at that point.

Consider an element of the tube at $P$, formed by two cross-sectional areas $\mathrm{d}S$, at right angles to the tube, separated by distances $\mathrm{d}l$. As before, we assume energy to be stored at a density $\frac{1}{2}DE$ per unit volume, so the energy stored in the element at $P$ is $\frac{1}{2}DE\mathrm{d}S\mathrm{d}l$. However, $D\mathrm{d}S$ is equal to $\mathrm{d}Q$ and $E\mathrm{d}l$ is equal to the potential difference $\mathrm{d}V$ between the faces of the element. Hence the energy stored in the element can be written as $\frac{1}{2}\mathrm{d}V\mathrm{d}Q$ and, for the whole tube, the stored energy is $\frac{1}{2}V\mathrm{d}Q$.

The whole of space can be filled with tubes of $D$ such as the one considered, so we conclude that the total stored energy is $\frac{1}{2}VQ$, where $Q$ is the magnitude of the charge on each conductor. This result is in agreement with (8.3) and we have shown that the assumption of stored energy $\frac{1}{2}DE$ per unit volume throughout the field gives the correct total stored energy for any pair of conductors in free space. By arguments similar to those employed in §3.7.6, it is a straightforward matter to extend the proposition to a system of any number of conductors, at different potentials, in free space.

Before making use of these results, we shall show that a similar procedure can be employed to calculate the energy stored as a result of the magnetic field caused by currents.

### 8.1.3 The energy stored when a current flows through an inductor

Let current $I$ be flowing through an inductor of inductance $L$ and, during an interval of time $\mathrm{d}t$, let the current be increased uniformly to $I+\mathrm{d}I$. This will cause an induced e.m.f. $L\,\mathrm{d}I/\mathrm{d}t$ to oppose the flow of current and during the interval $\mathrm{d}t$, the charge passing any point of the circuit is $I\mathrm{d}t$. Hence the work done against the opposing e.m.f. is

$$\left(L\frac{\mathrm{d}I}{\mathrm{d}t}\right) I\mathrm{d}t = LI\mathrm{d}I$$

and the total work done to establish a current $I$ in the inductor is

$$L\int_0^I I\mathrm{d}I = \tfrac{1}{2}LI^2 \tag{8.7}$$

This work represents stored energy, which can be recovered when the current is reduced to zero.

### 8.1.4 An alternative expression for the stored magnetic energy

We have already considered the energy stored in a magnetized isotropic toroidal specimen (§7.2.5), where the whole of the flux is confined to the specimen, and have shown that the energy per unit volume in this case amounts to

$$W_0 = \int_0^B H\mathrm{d}B$$

If the permeability $\mu$ of the specimen is constant,

$$W_0 = \frac{1}{\mu}\int_0^B B\mathrm{d}B = B^2/2\mu = \tfrac{1}{2}BH \tag{8.8}$$

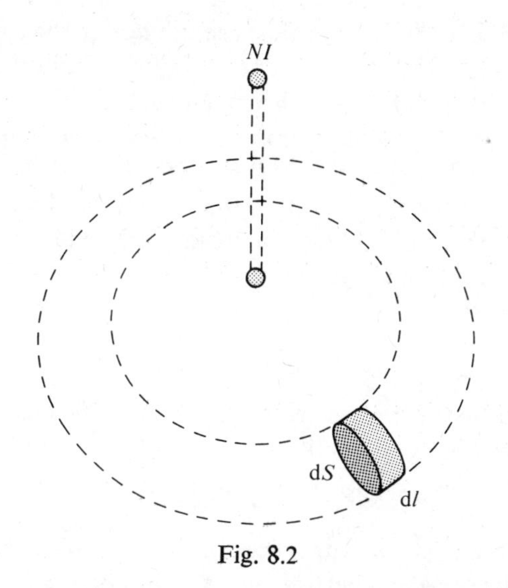


Fig. 8.2

We now wish to generalize this result and, for simplicity, shall consider a coil in free space having $N$ turns which are nearly coincident (fig. 8.2). Let current $I$ flow round the coil and let us fix our attention on a particular tube of magnetic flux which is generated by this current and which is completely linked with the coil. Let $\mathrm{d}\phi$ be the flux within this tube and consider an element of the tube, of length $\mathrm{d}l$ and area of cross-section $\mathrm{d}S$. If, in fact, energy is stored at the rate of $\frac{1}{2}BH$ per unit volume, the energy to be associated with the element is, since $H$ and $\mathrm{d}l$ are always parallel,

$$\frac{1}{2}\frac{\mathrm{d}\phi}{\mathrm{d}S}\,H\,\mathrm{d}l\,\mathrm{d}S = \tfrac{1}{2}\mathrm{d}\phi\; H\,\mathrm{d}l$$

and, for the whole tube, the energy is

$$\tfrac{1}{2}\mathrm{d}\phi \oint_l H\,\mathrm{d}l = \tfrac{1}{2}NI\,\mathrm{d}\phi \tag{8.9}$$

Suppose for the moment that the internal inductance of the coil is negligible and that we can assume the whole of the flux to be linked with all of the turns. Then the total flux linkage is

$$N\phi = LI \tag{8.10}$$

where $L$ is the inductance of the coil. But $\phi$ is made up of tubes of flux $\mathrm{d}\phi$ such as the one considered earlier. Thus, from (8.9) and (8.10), the total stored energy is $\frac{1}{2}LI^2$, which we know to be the correct value.

We have thus shown that an assumption of stored energy at the rate of

$\frac{1}{2}BH$ per unit volume throughout all space will lead to the correct value for the total energy stored when current flows through a coil. We do not assert that the energy is, in fact, stored throughout space.

When we considered the definition of $L$ (§4.5.3), we noted the complications arising from internal inductance and the fact that not all of the flux is linked with all of the turns or all of the current. We have avoided these complications in the above discussion because detailed investigation shows that they do not invalidate the final conclusion that we reached.

### 8.1.5 Extension of the above results

We have shown that, when electric and magnetic fields are established in istotropic media, the total energy stored can be calculated by assuming storage at the rate of

$$W_0 = \tfrac{1}{2}(DE+BH) \tag{8.11}$$

per unit volume throughout all space. If we deal with linear anisotropic materials, $\boldsymbol{D}$ will not be parallel to $\boldsymbol{E}$, nor $\boldsymbol{B}$ to $\boldsymbol{H}$. Equation (8.11) must then be modified to

$$W_0 = \tfrac{1}{2}(\boldsymbol{D}\cdot\boldsymbol{E}+\boldsymbol{B}\cdot\boldsymbol{H}) \tag{8.12}$$

Finally, if the material is non-linear, the equation becomes

$$W_0 = \int_0^D \boldsymbol{E}\cdot \mathrm{d}\boldsymbol{D}+\int_0^B \boldsymbol{H}\cdot \mathrm{d}\boldsymbol{B} \tag{8.13}$$

### 8.1.6 Energy storage in two coupled coils

In fig. 8.3 let currents $I_1$ and $I_2$ flow in coils 1 and 2, which have self-inductances $L_1$ and $L_2$ respectively. The permeability of the medium in which the coils are situated is constant; if it were not, $L_1$ and $L_2$ would have no precise meanings. There is magnetic coupling between the coils such that $I_2$ flowing in 2 produces flux linkage $M_{21}I_2$ with coil 1. Let the directions of $I_1$ and $I_2$ be such that this flux linkage reinforces that which $I_1$ produces in coil 1. We wish to find the total energy stored in this system.

We already know that the establishment of the two currents will lead to energy storage $\frac{1}{2}(L_1I_1^2+L_2I_2^2)$ by virtue of the self-inductances of the coils. However, there will now be additional energy storage as a result of the coupling between the coils. Suppose $I_1$ has been established in coil 1 and consider what takes place in this coil when the current in coil 2 is increased from $I$ to $I+\mathrm{d}I$ in time $\mathrm{d}t$. The change will cause e.m.f.

$$M_{21}\,\mathrm{d}I/\mathrm{d}t$$

to be induced in coil 1 in a direction opposing the flow of $I_1$. Moreover,

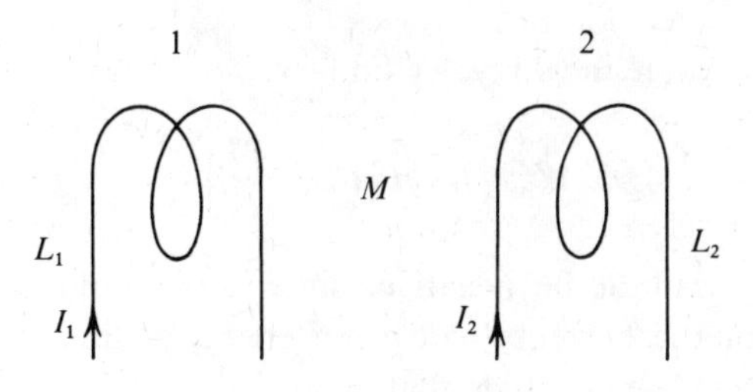


Fig. 8.3

in the interval $dt$, the charge flowing round coil 1 is $I_1 dt$, so the energy stored as a result of the induced e.m.f. is

$$M_{21} \frac{dI}{dt} I_1 dt = M_{21} I_1 dI$$

Thus, as a result of the coupling, the additional energy stored when the current in coil 2 is increased from 0 to $I_2$ is

$$M_{21} I_1 \int_0^{I_2} dI = M_{21} I_1 I_2$$

and the total energy of the system is

$$W_1 = \tfrac{1}{2}L_1 I_1^2 + \tfrac{1}{2}L_2 I_2^2 + M_{21} I_1 I_2$$

If we had started by establishing $I_2$ and had then increased the current in coil 1 from 0 to $I_1$, we should have found for the stored energy

$$W_2 = \tfrac{1}{2}L_1 I_1^2 + \tfrac{1}{2}L_2 I_2^2 + M_{12} I_1 I_2$$

Clearly $W_1$ and $W_2$ must be equal; if they were not we could store energy in the greater and release it in the lesser, to provide a constant source of energy. Thus

$$M_{21} = M_{12} = M \tag{8.14}$$

and we may speak of the mutual inductance of the coils without specifying in which the current is changing. We then have for the energy of the system

$$W = \tfrac{1}{2}L_1 I_1^2 + \tfrac{1}{2}L_2 I_2^2 + M I_1 I_2 \tag{8.15}$$

Let us now consider the same system, but with the direction of one of the currents reversed. The stored energy will then be

$$W' = \tfrac{1}{2}L_1 I_1^2 + \tfrac{1}{2}I_2 L_2^2 - M I_1 I_2 \tag{8.16}$$

Keeping $I_1$ fixed, let us calculate what value of $I_2$ will make $W'$ a minimum.

$$\frac{dW'}{dI_2} = L_2 I_2 - M I_1 = 0$$

or

$$I_2 = M I_1 / L_2$$

Substituting this value in (8.16), we find

$$W'_{\min} = \tfrac{1}{2}I_1^2\left(L_1 - \frac{M}{L_2}\right)$$

However, $W'_{\min}$ cannot be negative, since this would simply that the establishment of the currents had given energy to the sources of these currents. We therefore conclude that

$$M \ngtr \sqrt{(L_1 L_2)} \tag{8.17}$$

In practice $M$ may approach $\sqrt{(L_1 L_2)}$ very closely indeed. It is convenient to define a *coefficient of coupling* $k$ by the equation

$$k = M/\sqrt{(L_1 L_2)} \tag{8.18}$$

For $k = 1$, the coupling would be perfect.

## 8.2 Forces and couples

### 8.2.1 The force between charged parallel plates

The results of the previous section provide one of the best methods for calculating the forces between components of a system in which electric and/or magnetic fields exist. The general method is to assume some infinitesimal change to be made to the system and to equate the work done by external forces to the increase in energy of the system.

We begin by calculating the force per unit area between two large parallel plane conductors, one of which is charged to potential $V$ with respect to the other. Edge effects are to be neglected and we suppose the plates to be immersed in a fluid of permittivity $\epsilon$. The distance between the plates is $x$ (fig. 8.4($a$)). Let the lower plate be fixed and let the charges $+Q$ and $-Q$ per unit area respectively cause the top plate to be pulled downwards with a force $F$ per unit area. If $C$ is the capacitance and $W$ the stored energy, both per unit area,

$$C = \epsilon/x$$

and, from (8.2),

$$W = Q^2/2C$$

so

$$W = Q^2 x/2\epsilon \tag{8.19}$$

To determine $F$ we suppose $x$ to be increased by d$x$, but there are two sets of conditions under which this change may be made. First we assume the system to be isolated electrically, so that $Q$ remains constant and, since $C$ decreases, $V$ must increase. Equating the mechanical work done against

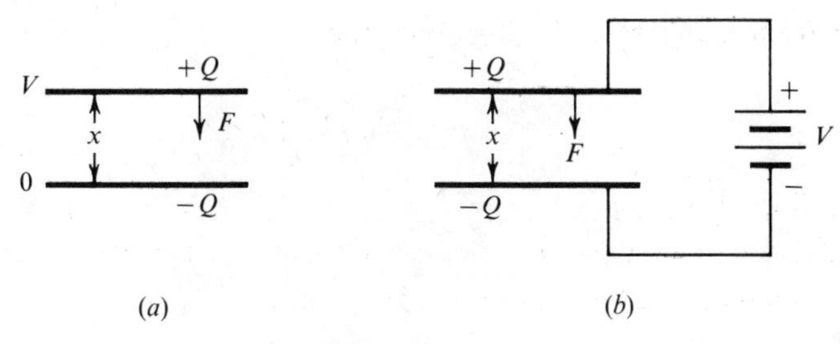


Fig. 8.4

$F$ to the increase in stored energy,

$$F\mathrm{d}x = Q^2\mathrm{d}x/2\epsilon$$

or
$$F = Q^2/2\epsilon = V^2C^2/2\epsilon = V^2\epsilon/2x^2 \tag{8.20}$$

It is instructive to consider the alternative procedure, (fig. 8.4(*b*)), in which the plates are attached to a battery so that $V$ remains constant during the change. In this case the decrease in $C$ causes $Q$ to decrease and the excess charge is forced back into the battery against the voltage $V$. If $\mathrm{d}Q$ is the increase in $Q$, we have

$$\underset{\text{(work done against } F)}{F\mathrm{d}x} + \underset{\text{(work against battery)}}{V\mathrm{d}Q} = \underset{\text{(decrease in stored energy)}}{\tfrac{1}{2}V\mathrm{d}Q}$$

or
$$F\mathrm{d}x = -\tfrac{1}{2}V\mathrm{d}Q \tag{8.21}$$

But
$$Q = VC = V\epsilon/x$$

Hence
$$\mathrm{d}Q = -V\epsilon\,\mathrm{d}x/x^2$$

and from (8.21)
$$F = V^2\epsilon/2x^2$$

in agreement with (8.20). Had we neglected the work done against the battery, we should have obtained a force of the right magnitude, but the wrong sign.

### 8.2.2 The electrostatic voltmeter

A simple form of electrostatic voltmeter (fig. 8.5) consists of a fixed set of parallel semi-circular plates $A$, interleaved with a similar rotatable set of plates $B$. A potential difference $V$ applied between the two sets produces a couple tending to increase the overlap of $B$ within $A$ and this couple is resisted by a torsion spring. The $B$ plates thus rotate until the two couples are equal.

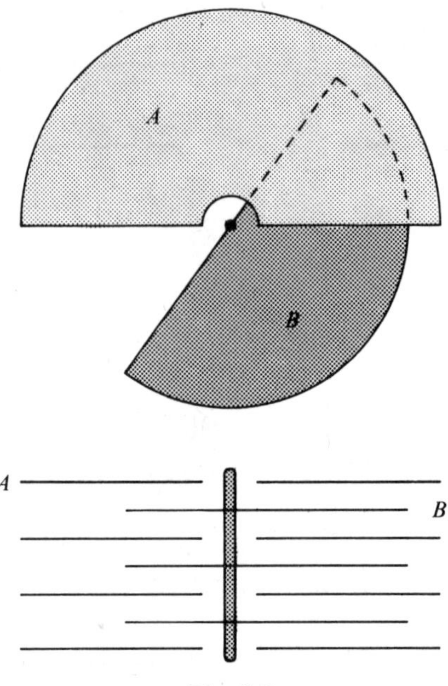


Fig. 8.5

If $k$ is the torsion constant of the spring, a deflection $\theta$ from the position of equilibrium when $V = 0$ will produce a couple

$$T_1 = k\theta \tag{8.22}$$

To calculate the couple $T_2$ resulting from connection of the plates to a potential difference $V$ we suppose that, after the plates have become charged, the system is electrically isolated, so that the charges $+Q$ and $-Q$ respectively, remain constant. If $C$ is the capacitance between the two sets of plates, the electrical stored energy is

$$W = Q^2/2C$$

If $\theta$ is increased by $\mathrm{d}\theta$, to give a corresponding increase $\mathrm{d}C$ in the capacitance, the change in the stored energy will be

$$\mathrm{d}W = -Q^2\mathrm{d}C/2C^2$$

This loss of energy must be equal to the work done by $T_2$, or

$$T_2\mathrm{d}\theta = Q^2\mathrm{d}C/2C^2$$

Thus

$$T_2 = \frac{1}{2}\frac{Q^2}{C^2}\frac{\mathrm{d}C}{\mathrm{d}\theta} = \tfrac{1}{2}V^2\frac{\mathrm{d}C}{\mathrm{d}\theta} \tag{8.23}$$

Equating $T_1$ and $T_2$

$$k\theta = \tfrac{1}{2}V^2\frac{\mathrm{d}C}{\mathrm{d}\theta} \tag{8.24}$$

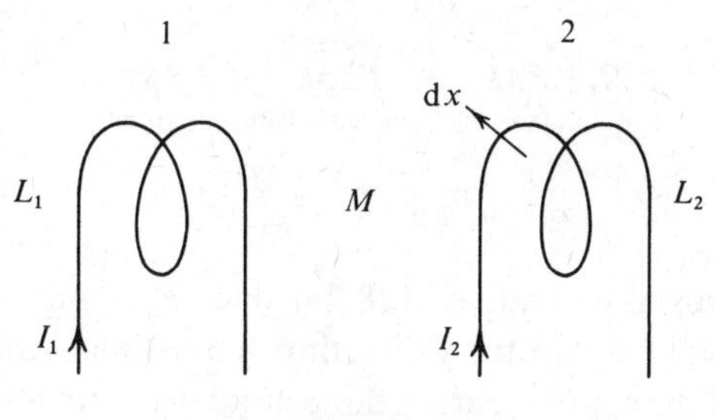


Fig. 8.6

The instrument is normally calibrated by the application of known voltages. Since the deflection is proportional to $V^2$, the meter can be used to measure both direct and alternating voltages. Depending on the shape of the rotatable plates, $dC/d\theta$ may or may not be independent of $\theta$.

### 8.2.3 Forces and couples between current-carrying coils

Consider two coils with self-inductances $L_1$ and $L_2$, carrying currents $I_1$ and $I_2$ respectively (fig. 8.6). If $M$ is their mutual inductance and if the directions of the currents are such that the flux linkages reinforce each other, we have seen that the total stored energy is

$$W = \tfrac{1}{2}L_1 I_1^2 + \tfrac{1}{2}L_2 I_2^2 + M I_1 I_2$$

We now suppose that, in time $dt$, coil 2 is given a small displacement $dx$ in some arbitrary direction, thereby increasing $M$ by $dM$, while $I_1$ and $I_2$ are kept constant. The stored energy will increase by $dM\, I_1 I_2$ and mechanical work $F_x dx$ will have been done, where $F_x$ is the component of force, in the direction of $dx$, experienced by coil 2 as a result of the interaction of $I_2$ with the flux produced by $I_1$.

Work will also have been done by the batteries which keep the currents constant. Taking coil 1, the e.m.f. induced by the change is $I_2 dM/dt$ and, during the interval $dt$, charge $I_1 dt$ flows round the coil.

Thus the work done is

$$dW = I_2 \frac{dM}{dt} I_1 dt = I_1 I_2 dM \tag{8.25}$$

The change increases the flux linkage with coil 1 and, by Lenz's law, the e.m.f. induced in this coil will be in a direction opposite to that of the current $I_1$, since we have postulated that the flux linkages from the two coils reinforce each other. We therefore conclude that the work $dW$ is done *by* the battery producing $I_1$. A similar argument shows that an equal amount of work is done by the battery producing $I_2$. Hence the overall energy balance is

$$\underset{\text{(from batteries)}}{2I_1I_2\mathrm{d}M} = \underset{\text{(mechanical)}}{F_x\mathrm{d}x} + \underset{\text{(stored)}}{I_1I_2\mathrm{d}M}$$

or

$$F_x = I_1I_2\frac{\partial M}{\partial x} \tag{8.26}$$

We use the partial derivative in (8.26) since $F_x$ is only one component of the total force, in an arbitrary direction. To calculate the total force, we might determine the components in three mutually perpendicular directions and then find their resultant. Alternatively, it is often possible to determine the direction of the total force from the symmetry of the system.

In (8.26) $F_x$ is the force acting on coil 2 in the direction of d$x$; that is, in the direction of increasing $M$. An equal and opposite force acts on coil 1.

Instead of the linear displacement of coil 2, we could have given it an angular displacement d$\theta$ about some arbitrary axis. By similar arguments we should then have found for the component of torque about this axis, acting on each coil,

$$T_x = I_1I_2\frac{\partial M}{\partial \theta} \tag{8.27}$$

The reader may find it strange that the force $F_x$ in (8.27) depends on the total flux $MI_1$ which is linked with coil 2, whereas an earlier discussion (§4.3.1) would indicate that the force is determined by the flux density $\boldsymbol{B}$ in the immediate vicinity of each element of $I_2$.

This apparent paradox can readily be explained by taking as the coils two coaxial single-turn circular loops, as shown in fig. 8.7, where the flux lines linked with coil 2 as a result of current $I_1$ in coil 1 are indicated. Round the circumference of coil 2 the flux density has a constant value $\boldsymbol{B}$ which can be resolved into an axial component $B_x$ and a radial component $B_r$. The former causes radial forces, which cancel for the whole coil, while the latter produces an axial force

$$F_x = 2\pi r_2 B_r I_2 \tag{8.28}$$

where $r_2$ is the radius of coil 2.

If coil 2 is given a small axial displacement d$x$, with consequent increase d$M$ in the mutual inductance, the additional flux linked with this coil will be

$$I_1\mathrm{d}M = 2\pi r_2 B_r \mathrm{d}x \tag{8.29}$$

From (8.28) and (8.29) we then have

$$F_x = I_1I_2\frac{\partial M}{\partial x}$$

as before. The argument can readily be extended to coils of more complicated shapes.

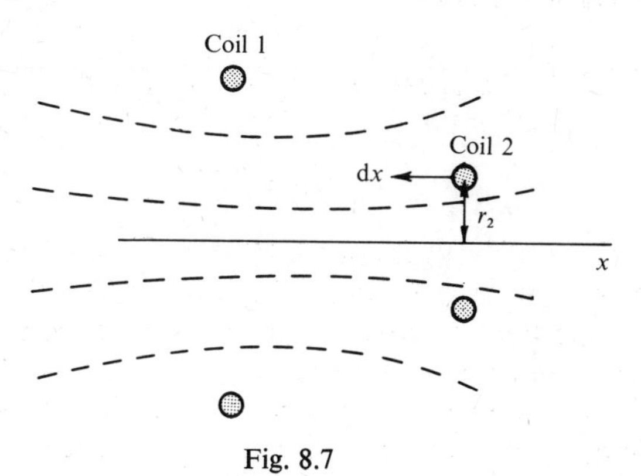


Fig. 8.7

### 8.2.4 Extension of the foregoing theory

The results obtained in the last section can be extended in various ways. For example, the force or torque acting on coil 2 is expressed in (8.26) and (8.27) in terms of $I_1$, $I_2$ and $M$. But $I_1 M$ is the flux linkage $\phi$ with coil 2, caused by current $I_1$ in coil 1. But, so far as coil 2 is concerned, neither the force nor the torque would be changed if the flux linkage were produced in some other way; for example by a permanent magnet. Thus it is convenient to write

$$\phi = MI_1, \quad \frac{\partial \phi}{\partial x} = I_1 \frac{\partial M}{\partial x}$$

whence

$$F_x = I_2 \frac{\partial \phi}{\partial \theta}, \quad T_\theta = I_2 \frac{\partial \phi}{\partial x} \tag{8.30}$$

where the components of force and torque acting on coil 2 are expressed in terms of the current flowing in this coil and the flux linkage with it, irrespective of the way in which this flux linkage has been caused.

### 8.2.5 Systems containing magnetic materials

Hitherto it has been tacitly assumed that the coils under consideration were in free space. However, in most practical devices where forces and torques are important, iron cores are generally used to concentrate the flux where it is needed. This circumstance does not in any way invalidate the arguments that we have used, so long as the fluxes are proportional to currents in the way that we have assumed. Even though iron itself is a non-linear material, this proportionality is usually a close approximation to the truth, since the iron circuit generally contains air gaps which keep its reluctance nearly constant. In what follows we shall assume that the system is linear.

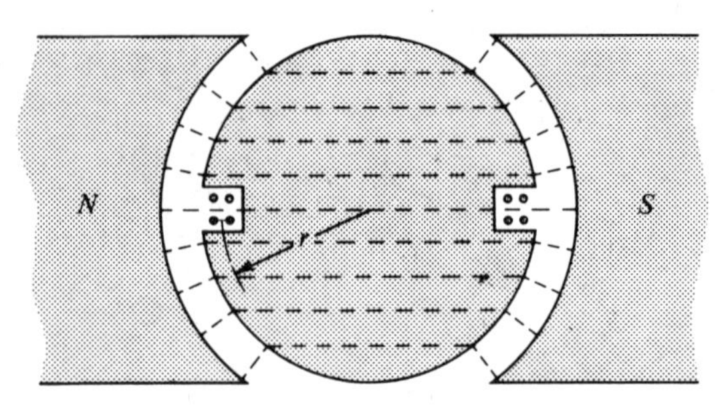


Fig. 8.8

An important practical problem is the calculation of the torque on an armature in an electrical machine. In fig. 8.8 we take a single rectangular coil of $N$ turns lying in the slots of an armature. The width of the coil is $2r$, its axial length is $l$ and it carries current $I$. The field magnet provides radial flux density $B$ in the air gap.

If the armature is given a small angular deflection $d\theta$, the change in the flux linkage is

$$d\phi = 2NBrl\,d\theta$$

Hence, from (8.30) the torque is

$$T_\theta = 2NBIrl \tag{8.31}$$

This is exactly the torque that would have been obtained if each axial wire of the coil had lain in a radial field of flux density $B$. Although this latter method of calculation is often erroneously used, in fact the coil is largely shielded from the magnetic field by the iron, through which the flux passes. The torque is thus exerted on the iron of the armature and only to a very small extent on the coil itself.

The presence of iron or other magnetic material in the field gives rise to another possibility, which was not envisaged in the treatment of the force between two coils in free space. This is, that motion of the iron, by re-distributing the flux, may change not only the mutual inductance, but also the self-inductance of the coils. We consider first the force on a single coil when such a change occurs.

Let a coil of self-inductance $L$, carrying current $I$ which is kept constant, be given a small displacement $dx$ in any arbitrary direction, such that the inductance increases to $L+dL$. The increase in the stored energy is $\frac{1}{2}I^2dL$ and the change in the flux linkage $IdL$. If the change takes place in time $dt$, the induced e.m.f. will be $IdL/dt$, in a direction which opposes the change. During this interval the charge carried round the circuit is $Idt$, so the battery supplying $I$ does additional work

$$dW = I\frac{dL}{dt}Idt = I^2dL$$

to keep the current constant. If $F_x$ is the mechanical force in the direction $dx$, the energy balance gives

$$\underset{\text{(from battery)}}{I^2dL} = \underset{\text{(stored energy)}}{\tfrac{1}{2}I^2dL} + \underset{\text{(mechanical work)}}{F_x dx}$$

or

$$F_x = \tfrac{1}{2}I^2\frac{\partial L}{\partial x} \tag{8.32}$$

We next consider the case of two coils, with inductances $L_1$ and $L_2$ and carrying currents $I_1$ and $I_2$ respectively. The mutual inductance between the coils is $M$ and we suppose that any change in the system alters $L_1$ and $L_2$ as well as $M$. By arguments exactly similar to those already employed, the total component of force between the coils is found to be

$$F_x = \tfrac{1}{2}I_1^2\frac{\partial L_1}{\partial x}+\tfrac{1}{2}I_2^2\frac{\partial L_2}{\partial x}+I_1I_2\frac{\partial M}{\partial x} \tag{8.33}$$

and similarly

$$T_\theta = \tfrac{1}{2}I_1^2\frac{\partial L_1}{\partial \theta}+\tfrac{1}{2}I_2^2\frac{\partial L_2}{\partial \theta}+I_1I_2\frac{\partial M}{\partial \theta} \tag{8.34}$$

These equations reduce to (8.26) and (8.27) when $L_1$ and $L_2$ remain constant. They give us no information as to where the force $F_x$ and the couple $T_\theta$ act. The action may be on the coils themselves or on any magnetic material involved in the change, or it may be partly on each.

### 8.2.6 The force between magnetized pole pieces

Fig. 8.9 shows an electromagnet with a winding of $N$ turns carrying current $I$ and producing flux density $B$ round the magnetic circuit. There are two air gaps, each of width $x$, and it is assumed that fringing of the flux can be neglected. It is also assumed that $x$ is large enough to control the reluctance of the magnetic circuit, so that the system is linear and $B$ is proportional to $I$.

The area of cross-section of the magnetic material is $A$ and we wish to find the force $F$ between the two halves of the magnet. We do this by considering the energy balance when $x$ is increased by $dx$.

We first assume that, as $x$ is increased, $I$ is simultaneously increased to keep $B$ constant. Then no e.m.f. will be induced in the winding and any additional work done by the battery will appear as heat in the resistance of the circuit, and so may be ignored. The energy stored in the magnetic

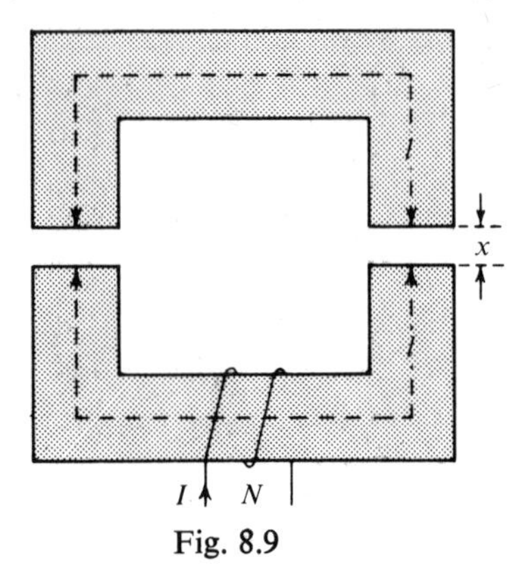


Fig. 8.9

material will be unchanged, so the work done against the force $F$ must be equal to the increased energy stored in the gap. Thus

$$F\,dx + (B^2/2\mu_0)2A\,dx = 0$$

whence
$$F = -B^2A/\mu_0 \tag{8.35}$$

or, the force $F_0$ per unit area of gap is

$$F_0 = -B^2/2\mu_0 \tag{8.36}$$

The negative sign indicates that $F$ is in the opposite direction to $dx$; that is, it is a force of attraction between the two halves.

We now seek to derive this force from (8.32). The flux $\phi$ round the magnetic circuit can be written as

$$\phi = NI/[(2x/\mu_0 A) + R] \tag{8.37}$$

where the first term in the denominator is the reluctance of the two air gaps and $R$ is the reluctance of the magnetic material. With our assumption of linearity, $R$ is a constant. Then

$$L = N\phi/I = \mu_0 N^2 A/(2x + \mu_0 RA)$$

and
$$F_x = \tfrac{1}{2}I^2 \frac{\partial L}{\partial x} = -\mu_0 N^2 A I^2/(2x + \mu_0 RA)^2$$

which, from (8.37), becomes

$$F_x = -\phi^2/\mu_0 A = -B^2A/\mu_0$$

in agreement with (8.35).

At this point the reader may wonder why it has been necessary to insist on the linearity of the system. The force between the two halves of the magnet depends on the flux density $B$ in a particular state of the material and surely cannot have been influenced by the relation between $B$ and $H$ while that state was being reached. This is quite true and the necessity for stipulating linearity is that self- and mutual-inductances have no precise meaning in a non-linear system. On the other hand, it is quite legitimate to use the energy-balance method to derive forces and couples when non-linear materials are involved, provided we remember that the energy per unit volume stored in such material is not $\frac{1}{2}BH$ but $\int_0^B H\,dB$. To illustrate this point, let us suppose the two halves of the system of fig. 8.9 to be permanent magnets and the current $I$ to be zero.

Let $H_0$ be the magnetic field in the air gaps and $H_1$ its value in the magnetic material. If $2l$ is the total length of path in the material, the line integral of $H$ round the magnetic circuit is

$$2(xH_0+lH_1) = 0 \tag{8.38}$$

When the air gaps are increased to $x+dx$, the fields become $H_0+dH_0$ and $H_1+dH_1$ respectively and $B$ changes to $B+dB$.

The energy stored in the air gaps (which are linear) increases by

$$2A[\tfrac{1}{2}(B+dB)(H_0+dH_0)(x+dx)-\tfrac{1}{2}BH_0x]$$

Retaining only small quantities of the first order, and remembering that

$$B = \mu_0 H, \quad dB = \mu_0 dH$$

this becomes
$$A(2H_0x\,dB+BH_0\,dx)$$

The increase in energy stored in the material is $2AlH_1\,dB$ and, from (8.38), this is equal to $-2AH_0x\,dB$. Thus, for the overall energy balance, we have

$$F\,dx+ABH_0\,dx = 0$$

or
$$F = -ABH_0 = -AB^2/\mu_0$$

in agreement with (8.35).

The reader may care to check that, if stored energy at the rate of $\frac{1}{2}BH$ had been assumed for both the air gaps and the material, $F$ would have been found to be zero. This is in agreement with the fact that in a linear system there would be no hysteresis and hence no permanent magnetism.

## 8.3 Worked example

The inner conductor of a coaxial cable has radius $r_1$ and the internal radius of the thin external conductor is $r_2$. Derive an expression for the magnetic energy stored per unit length of cable when current $I$ flows through it. Hence determine the inductance per unit length of cable.

*Solution.* From §4.5.4, the circumferential magnetic field $H_r$, at radius $r$, has the following values:

$$H_r = Ir/2\pi r_1^2 \quad \text{for } r < r_1$$

$$H_r = I/2\pi r \quad \text{for } r_1 < r < r_2$$

Assuming $\mu_r$ for the materials of the cable to be equal to unity, the magnetic energy stored per unit volume is $\frac{1}{2}\mu_0 H^2$. Thus, the energy per unit length stored between $r$ and $r+\mathrm{d}r$ is

$$\tfrac{1}{2}\mu_0 H_r^2\, 2\pi r\,\mathrm{d}r = \pi\mu_0 r H_r^2\,\mathrm{d}r$$

and the total energy per unit length is

$$\frac{\mu_0 I^2}{4\pi}\left(\frac{1}{r_1}\int_0^{r_1} r^3\,\mathrm{d}r + \int_{r_1}^{r_2}\frac{\mathrm{d}r}{r}\right) = \frac{\mu_0 I^2}{4\pi}\left(\tfrac{1}{4}+\ln\frac{r_2}{r_1}\right)$$

But this is equal to $\frac{1}{2}LI^2$, if $L$ is the inductance per unit length of cable Hence

$$L = \frac{\mu_0}{2\pi}\left(\tfrac{1}{4}+\ln\frac{r_2}{r_1}\right)$$

in agreement with (4.114).

## 8.4 Problems

1. Show that the maximum energy that can be stored in a parallel-plate capacitor is $\epsilon_r\epsilon_0 b^2/2$ per unit volume, where $b$ is the dielectric strength of the insulator (maximum electric field strength before break-down) and $\epsilon_r\epsilon_0$ is its permittivity.

Make a rough comparison between this maximum energy and the energy stored per unit volume of a lead–acid storage battery, taking as typical values $\epsilon_r = 3$, $b = 1.5\times10^6$ V cm$^{-1}$.

2. A rectangular block of dielectric material, of relative permittivity $\epsilon_r$, is partially inserted between two much larger parallel plane conducting plates, as in fig. 8.10. If $d$ is the distance between the plates and $V$ the potential difference between them, find an expression for the pressure $P$ on the face shown in the figure.

3. Calculate the force between coaxial circular coils, distant $x$ apart, when they have radii $R_1$ and $R_2$, numbers of turns $N_1$ and $N_2$ and carry currents $I_1$ and $I_2$ respectively. The currents flow in the same directions in the two coils and it may be assumed that $R_1$ is large compared with $R_2$ and with $x$.

4. A single plane loop of wire, of area $A$ and any shape, carries current $I$. It lies in a uniform magnetic field, of flux density $B$, with the normal to its plane making angle $\theta$ with the direction of the field. Derive an expression for the magnitude of the torque acting on the loop.

5. A small piece of paramagnetic material, of volume $v$ and constant relative permeability $\mu_r$, lies in a non-uniform magnetic field. It may be assumed that $\mu_r$

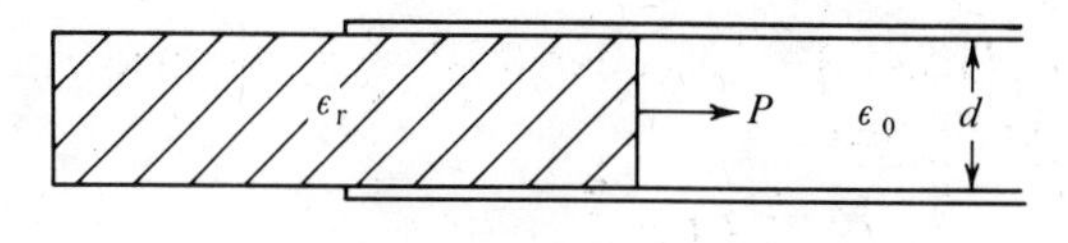


Fig. 8.10

is nearly equal to unity, so that the magnetic field inside the material is the same as that outside.

Show that the force acting on the material in any direction $x$ is

$$F_x = \frac{v\mu_0}{2}(\mu_r - 1)\frac{\partial H^2}{\partial x}$$

Hence show that, if a rod of the material of cross-section $A$ hangs vertically with its lower end in a uniform horizontal field $H_1$, while its upper end is in a uniform horizontal field $H_2$, the downward vertical magnetic force on the rod is

$$F = \tfrac{1}{2}\mu_0 A(\mu_r - 1)(H_1^2 - H_2^2)$$

(N.B. Each of the above equations forms the basis of a method of measuring the permeability of a weakly magnetic material.)

6. In a moving-iron ammeter, deflection of the needle causes the inductance of the winding to change in accordance with the relation

$$L = b + f(\theta)$$

where $b$ is a constant and $\theta$ is the angular deflection. The restoring torque on the needle is proportional to $\theta$.

What form of the function $f(\theta)$ would cause the deflection to be proportional to the four-thirds power of the current flowing through the winding?

If this form could be achieved, would the meter then read r.m.s. values of alternating current when the waveform was not sinusoidal?

# 9

## Electromagnetic waves

### 9.1 Summary of equations previously derived

In our discussion of the electrostatic field in free space we showed that Gauss' theorem, following directly from the inverse-square law led to the equation (3.28)

$$\oint_S \boldsymbol{D}\cdot\boldsymbol{n}\,\mathrm{d}S = \Sigma Q \tag{9.1}$$

It was later shown (§5.3.2) that this equation holds also when any number of homogeneous, isotropic dielectric media are present in the field.

For the flow of current in a material medium, we have the similar equation (2.28)

$$\oint_S \boldsymbol{J}\cdot\boldsymbol{n}\,\mathrm{d}S = \Sigma I \tag{9.2}$$

The treatment of the magnetic field in free space was based on the experimental relation (4.6)

$$\boldsymbol{B} = \frac{\mu_0 I}{4\pi}\oint_l \frac{\mathrm{d}\boldsymbol{l}\times\boldsymbol{r}_0}{r^2} \tag{9.3}$$

and, from this it was shown (§4.1.9) that $\boldsymbol{B}$ is a flux vector. Experimental evidence tells us that single magnetic poles do not exist and it is therefore legitimate to ascribe the production of all magnetic fields to currents (§4.1.1), as is done in (9.3). It then follows that lines of magnetic flux form closed loops, so we may write (4.8)

$$\oint_S \boldsymbol{B}\cdot\boldsymbol{n}\,\mathrm{d}S = 0 \tag{9.4}$$

A further derivation from (9.3) was expressed in the relation (4.50)

$$\oint_S \boldsymbol{H}\cdot\mathrm{d}\boldsymbol{s} = I \tag{9.5}$$

In §5.4.2 it was shown that (9.4) and (9.5) are also true for magnetic fields containing any number of homogeneous isotropic materials.

Our discussion of electromagnetic induction was based on Faraday's law (4.90) and it was shown that, when the flux linkage with a circuit changes with time, both as a result of change in $\boldsymbol{B}$ and also because elements of the circuit are moving through the field, this law can be expressed in the form (4.91)

$$\text{e.m.f.} = \oint_s (\boldsymbol{u}\times\boldsymbol{B})\cdot d\boldsymbol{s} - \int_S \frac{\partial \boldsymbol{B}}{\partial t}\cdot\boldsymbol{n}\,dS \tag{9.6}$$

When all parts of the circuit are at rest, relative to the observer, the first term on the right-hand side of (9.6) vanishes, and this equation becomes

$$\text{e.m.f.} = -\int_S \frac{\partial \boldsymbol{B}}{\partial t}\cdot\boldsymbol{n}\,dS \tag{9.7}$$

Since $\boldsymbol{B}$ is a flux vector in the presence of magnetic materials, (9.7) will remain true when these materials are present in the field.

The above equations are all expressed in integral form and it was shown in chapter 6 that they lead to differential forms, relating to conditions at a particular point of the field. Thus (9.1) gives (6.4)

$$\operatorname{div}\boldsymbol{D} = \nabla\cdot\boldsymbol{D} = \rho \tag{9.8}$$

(9.4) gives (6.10)

$$\operatorname{div}\boldsymbol{B} = \nabla\cdot\boldsymbol{B} = 0 \tag{9.9}$$

(9.5) gives (6.90)

$$\operatorname{curl}\boldsymbol{H} = \nabla\times\boldsymbol{H} = \boldsymbol{J} \tag{9.10}$$

and (9.7) gives (6.95)

$$\operatorname{curl}\boldsymbol{E} = \nabla\times\boldsymbol{E} = -\frac{\partial \boldsymbol{B}}{\partial t} \tag{9.11}$$

In addition, we have by definition

$$\boldsymbol{D} = \epsilon\boldsymbol{E}, \quad \boldsymbol{B} = \mu\boldsymbol{H}, \quad \boldsymbol{J} = \sigma\boldsymbol{E} \tag{9.12}$$

Finally, in a conducting medium where currents are changing with time, the indestructibility of charge leads to the conclusion that the total outflow of current through a closed surface must be equal to the rate at which the total charge within the surface is decreasing.

Thus (9.2) may be expressed as

$$\oint_S \boldsymbol{J}\cdot\boldsymbol{n}\,dS = -\int_v \frac{\partial\rho}{\partial t}\,dv$$

giving finally

$$\operatorname{div}\boldsymbol{J} = \nabla\cdot\boldsymbol{J} = -\frac{\partial\rho}{\partial t} \tag{9.13}$$

## 9.2 The Maxwell hypothesis

Reviewing the above questions, Maxwell observed that, in certain situations, (9.5) and (9.10) lead to an anomaly.

In fig. 9.1, $C$ is a parallel-plate capacitor forming part of a circuit in which current $I$ is flowing as a result of the finite conductivity of the

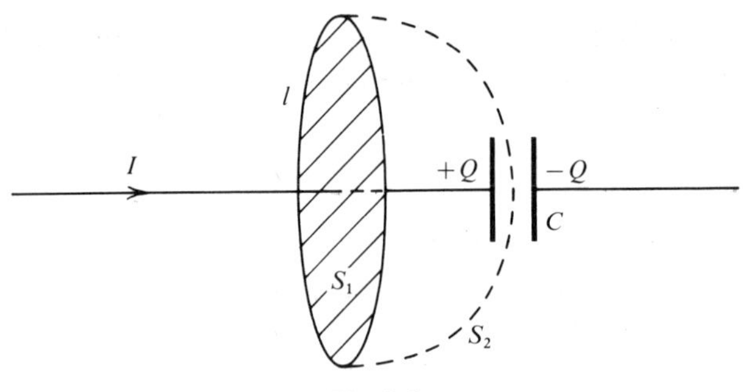


Fig. 9.1

dielectric between the plates of $C$. $S_1$ is a surface, with boundary $l$, lying wholly outside the capacitor $C$. $S_2$ is another surface, with the same boundary $l$, but passing between the plates of $C$.

If conditions are not changing with time, $I$ will be constant. Outside the capacitor, $I$ will flow along the wire of the circuit; within $C$ it will flow through the dielectric, which has finite conductivity. For either of the surfaces $S_1$ and $S_2$ we can write

$$\oint \boldsymbol{H}\cdot \mathrm{d}\boldsymbol{l} = I = \int_S \boldsymbol{J}\cdot\boldsymbol{n}\,\mathrm{d}S \tag{9.14}$$

If, however, as a result of a changing e.m.f. in the rest of the circuit, $I$ is varying with time, the current flowing along the wire will no longer be equal to the *conduction* current through the dielectric of $C$. The total current along the wire will be made up of a component resulting from the conductivity of the dielectric and an additional component $\mathrm{d}Q/\mathrm{d}t$ caused by the varying charges, $+Q$ and $-Q$ respectively, on the plates of $C$. Remembering that unit flux of $\boldsymbol{D}$ originates from unit charge, we may write

$$I = \int_S \boldsymbol{J}\cdot\boldsymbol{n}\,\mathrm{d}S + \int_S \frac{\partial \boldsymbol{D}}{\partial t}\cdot\boldsymbol{n}\,\mathrm{d}S \tag{9.15}$$

and this will be true for each of the surfaces $S_1$ and $S_2$.

Hitherto we have tacitly assumed that magnetic fields are produced only by conduction currents, but it now appears that, on this basis, (9.14) cannot be true for both of the surfaces $S_1$ and $S_2$. Maxwell suggested that this difficulty would disappear if we assume that both terms on the right-hand side of (9.15) are effective in producing magnetic fields. The first term is the conduction current and the second term is known as the *displacement current*.

There is no direct proof of the validity of Maxwell's hypothesis, but the indirect evidence is overwhelming and the hypothesis is now accepted as one of the fundamental laws of electromagnetic theory. Some of the evidence in support of it will be described in the next section.

## 9.3 The wave equation

### 9.3.1 The general equation

Modifying our previous summary of the field equations to take account of Maxwell's hypothesis, we have

$$\nabla \cdot \boldsymbol{D} = \rho \tag{9.16}$$

$$\nabla \cdot \boldsymbol{B} = 0 \tag{9.17}$$

$$\nabla \times \boldsymbol{E} = -\frac{\partial \boldsymbol{B}}{\partial t} \tag{9.18}$$

$$\nabla \times \boldsymbol{H} = \boldsymbol{J} + \frac{\partial \boldsymbol{D}}{\partial t} \tag{9.19}$$

which are generally known as Maxwell's equations. We shall now show that mathematical manipulation of these equations leads to a wave equation. For simplicity, our derivation will be limited to a single medium, with constant permittivity $\epsilon$ and permeability $\mu$ and with zero conductivity, so that (9.19) becomes

$$\nabla \times \boldsymbol{H} = \frac{\partial \boldsymbol{D}}{\partial t} \tag{9.20}$$

We further suppose that there is no distributed charge $\rho$, so that (9.16) becomes

$$\nabla \cdot \boldsymbol{D} = \mu \nabla \cdot \boldsymbol{E} = 0 \tag{9.21}$$

We begin by taking the curl of both sides of (9.18) to give

$$\nabla \times (\nabla \times \boldsymbol{E}) = -\nabla \times \left(\frac{\partial \boldsymbol{B}}{\partial t}\right) \tag{9.22}$$

From an identity previously proved (6.108), this can be written

$$\nabla(\nabla \cdot \boldsymbol{E}) - \nabla^2 \boldsymbol{E} = -\nabla \times \left(\frac{\partial \boldsymbol{B}}{\partial t}\right)$$

Substituting from (9.21) this becomes

$$\nabla^2 \boldsymbol{E} = \nabla \times \left(\frac{\partial \boldsymbol{B}}{\partial t}\right) \tag{9.23}$$

Changing the order of the partial derivative, we have, using (9.20)

$$\nabla \times \left(\frac{\partial \boldsymbol{B}}{\partial t}\right) = \frac{\partial}{\partial t}(\nabla \times \boldsymbol{B}) = \mu \frac{\partial}{\partial t}(\nabla \times \boldsymbol{H}) = \mu \frac{\partial^2 \boldsymbol{D}}{\partial t^2} = \mu\epsilon \frac{\partial^2 \boldsymbol{E}}{\partial t^2} \tag{9.24}$$

Thus, from (9.23) and (9.24) $\nabla^2 \boldsymbol{E} = \mu\epsilon \dfrac{\partial^2 \boldsymbol{E}}{\partial t^2}$ (9.25)

This is the well known three-dimensional wave equation, which tells us that any change in $\boldsymbol{E}$ at a particular point is not observed instantaneously at other points, but is propagated through the medium with finite velocity $v$, where

$$v = 1/\sqrt{(\mu\epsilon)} \tag{9.26}$$

A somewhat similar procedure leads to an analogous wave equation for $\boldsymbol{B}$, so magnetic changes, also, are propagated through the medium with velocity $v$. We have to accept that changes in the electromagnetic field cannot be established instantaneously; they are propagated outwards with velocity $v$ from whatever agencies are causing the changes.

### 9.3.2 Plane electromagnetic waves

To examine the properties of electromagnetic waves in greater detail it will be convenient to restrict our analysis to the one dimensional case. This means that, if the wave is travelling in, say, the $x$-direction, the values of $\boldsymbol{E}$, $\boldsymbol{D}$, $\boldsymbol{B}$ and $\boldsymbol{H}$ at any instant are the same at all points in a plane at right angles to the $x$-axis. A wave of this type is known as a plane wave and it is not one which normally occurs in practice. As a rule, if a change in the field is caused at a particular place, it is propagated outwards in all directions. If the propagation were uniform in all directions we should have a spherical wave; that is, the locus of all points which, at any given instant, had identical values of $\boldsymbol{E}$, $\boldsymbol{D}$, $\boldsymbol{B}$ and $\boldsymbol{H}$, would be a sphere. At a great distance from the source, a small portion of the spherical wave would appear to an observer to be very nearly a plane wave and it is an idealized wave of this type that we are considering.

The restriction to a plane wave travelling in the $x$-direction means that the partial derivatives of all quantities with respect to $y$ and $z$ must be put equal to zero.

We then find from (9.17) and (9.21) that

$$\frac{\partial D_x}{\partial x} = \frac{\partial E_x}{\partial x} = \frac{\partial B_x}{\partial x} = \frac{\partial H_x}{\partial x} = 0 \tag{9.27}$$

Similarly, if we expand (9.18) to give

$$\boldsymbol{i}\left(\frac{\partial E_z}{\partial y} - \frac{\partial E_y}{\partial z}\right) + \boldsymbol{j}\left(\frac{\partial E_x}{\partial z} - \frac{\partial E_z}{\partial x}\right) + \boldsymbol{k}\left(\frac{\partial E_y}{\partial x} - \frac{\partial E_x}{\partial y}\right) = -\frac{\partial}{\partial t}(\boldsymbol{i}B_x + \boldsymbol{j}B_y + \boldsymbol{k}B_z)$$

and equate corresponding components, we find

$$\frac{\partial B_x}{\partial t} = \frac{\partial H_x}{\partial t} = 0 \tag{9.28}$$

$$\frac{\partial B_y}{\partial t} = \mu \frac{\partial H_y}{\partial t} = \frac{\partial E_z}{\partial x} \tag{9.29}$$

$$\frac{\partial B_z}{\partial t} = \mu \frac{\partial H_z}{\partial t} = -\frac{\partial E_y}{\partial x} \tag{9.30}$$

Similarly treatment of (9.20) gives

$$\frac{\partial D_x}{\partial t} = \frac{\partial E_x}{\partial t} = 0 \tag{9.31}$$

$$\frac{\partial D_y}{\partial t} = \epsilon \frac{\partial E_y}{\partial t} = -\frac{\partial H_z}{\partial x} \tag{9.32}$$

$$\frac{\partial D_z}{\partial t} = \epsilon \frac{\partial E_z}{\partial t} = \frac{\partial H_y}{\partial x} \tag{9.33}$$

From (9.27), (9.28) and (9.31) it appears that the $x$-components of $\boldsymbol{E}$ and $\boldsymbol{H}$ cannot vary with respect to either $x$ or $t$. The component in the $x$-direction can only represent a uniform steady field. This is not a wave and is of no interest to us; we shall not consider it further. The electric and magnetic components of the wave are entirely transverse with respect to the direction of propagation.

From (9.29), (9.30), (9.32) and (9.33) we see that the $z$-component of $\boldsymbol{E}$ is associated with the $y$-component of $\boldsymbol{H}$ in its variation with respect to both $x$ and $t$. Similarly, the $y$-component of $\boldsymbol{E}$ is associated with the $z$-component of $\boldsymbol{H}$. We thus have an overall picture of two separate waves, in each of which the $\boldsymbol{E}$-component is at right angles to the $\boldsymbol{H}$ component. For our present purpose it will be sufficient to consider only one of these waves and we suppose the directions of the axes to be such that $E_z$ and $H_y$ are both zero. In electrical theory such a wave would be said to be *polarized in the xy-plane* but, in text books on optics, for historical reasons, the plane of polarization is commonly taken to be that in which the magnetic vector lies.

With the various restrictions that we have imposed, (9.25) becomes

$$\frac{\partial^2 E_y}{\partial x^2} = \frac{1}{v^2}\frac{\partial^2 E_y}{\partial t^2} \tag{9.34}$$

where $$v = 1/\sqrt{(\mu\epsilon)}$$

A general solution of (9.34) is of the form

$$E_y = f_1(x - vt) + f_2(x + vt) \tag{9.35}$$

where $f_1$ and $f_2$ are functions of any form. The first term on the right-hand side represents a wave travelling with velocity $v$ in the direction of $x$

increasing, while the second term is a wave travelling with the same velocity in the opposite direction.

To find the magnetic component of the wave, we use (9.32) and (9.35) to give

$$\frac{\partial H_z}{\partial x} = -\epsilon \frac{\partial E_y}{\partial t} = \epsilon v\,[f_1'(x-vt)-f_2'(x+vt)] \tag{9.36}$$

Integrating, and remembering that $v = 1/\sqrt{(\mu\epsilon)}$

$$H_z = \left(\sqrt{\frac{\epsilon}{\mu}}\right)[f_1(x-vt)-f_2(x+vt)] \tag{9.37}$$

Thus, for the wave travelling in the direction of $x$ increasing, the ratio of $H_z$ to $E_y$ has the constant value $\sqrt{(\epsilon/\mu)}$; for the wave in the opposite direction, the ratio has the same value but opposite sign.

Since $H_z$ is measured in amperes per metre, while $E_y$ is in volts per metre, the ratio $E_y/H_z$ has the dimensions of a resistance or impedance. It is often known as the *intrinsic impedance* $Z_0$ of the medium. Thus

$$Z_0 = E_y/H_z = \sqrt{(\mu/\epsilon)} = \mu v \tag{9.38}$$

For a plane wave in free space its value is approximately 376.7 ohms.

It should be emphasized that the electric and magnetic components are essential constituents of any electromagnetic wave and one cannot exist without the other. In fact, by virtue of (9.18) and (9.20), it is variation of each with time that gives rise to the other.

### 9.3.3 Light as an electromagnetic wave

In Maxwell's time it was accepted that visible light travels through space as a wave and Maxwell's own equations indicated that electromagnetic disturbances should also be propagated as transverse waves. He was therefore led to consider an earlier speculation by Faraday that light might be some form of electromagnetic disturbance. The velocity of light in free space was known to be about $3 \times 10^8$ m s$^{-1}$ and, from measurements made by himself and others, Maxwell concluded that $1/\sqrt{(\mu_0\epsilon_0)}$ would have almost exactly this same value. This was sufficient to convince him that light was, in fact, an electromagnetic wave though, at the time, this conclusion was not universally accepted. However, some twenty-five years later, Hertz showed that waves exhibiting the usual phenomena of reflection, refraction and interference could be produced by purely electromagnetic means, and the truth of Maxwell's theory was then no longer in doubt.

## 9.4 The quasi-stationary state

In the earlier chapters of this book, essential conclusions from Maxwell's theory – the finite velocity of propagation of electric or magnetic changes and the production of magnetic fields by displacement currents – were completely ignored. It is pertinent to enquire how far this neglect invalidates the theory which we established.

The answer to this question depends on the linear dimensions of the system that we are considering. If the time taken for an electromagnetic disturbance to be propagated from one end of the system to the other is always very small compared with the time occupied by any change that we are observing, the Maxwell effects can safely be ignored: otherwise they must be considered.

If we think in terms of alternating-current experiments on a laboratory scale, Maxwell effects are unlikely to be important at frequencies below a few tens of kilohertz, unless we are concerned with the propagation of electrical or magnetic changes through materials with very high values of $\epsilon$ or $\mu$. On the other hand, if we were dealing with a long-wave aerial system, the effects could certainly not be neglected at these frequencies. For the transmission of power across a continent, the velocity of propagation is important at 50 Hz.

Changes in which the Maxwell effects can be ignored are often said to be *quasi-stationary*.

# 10

# The experimental basis of electromagnetic theory and some applications

## 10.1 The experimental evidence on which electromagnetic theory rests

### 10.1.1 Introduction

At various points throughout this book, statements have been made which can be justified only by appeal to experimental results. For some of these statements (e.g. that single free magnetic poles do not exist) one cannot point to any experiment or group of experiments that provide adequate justification. One can only say that, if such statements were not true, phenomena demonstrating their falsity would have been observed. Furthermore, until phenomena of this kind are observed, our theory is adequate to account for all experimental facts and that is all that can be asked of any theory.

Apart from these rather general appeals to experience, there are other cases where we have based our theory on exact quantitative laws (e.g. Coulomb's inverse-square law of force between charges), which can only be verified by experiment. In this section we shall briefly review some of the experimental evidence and shall indicate the accuracy with which the relevant measurements have been carried out.

### 10.1.2 The inverse-square law of force between charges

No phenomena have ever been observed to indicate that the force between point charges does not act along the line joining them, so our problem is to decide whether the force is inversely proportional to the square of the distance separating them. We have shown that the inverse square law leads to the conclusion that, as a result of external charges, there can be no field inside a closed conductor (§3.6.4). We now prove that if the law of force were

$$\boldsymbol{F} = Q_1 Q_2 / 4\pi\epsilon_0 r^n \quad \text{where} \quad n \neq 2 \tag{10.1}$$

the field would not be zero inside a spherical conductor.

In fig. 10.1(*a*) we suppose the spherical conductor to have uniform charge $Q$ per unit area, and we consider an elemental conical surface

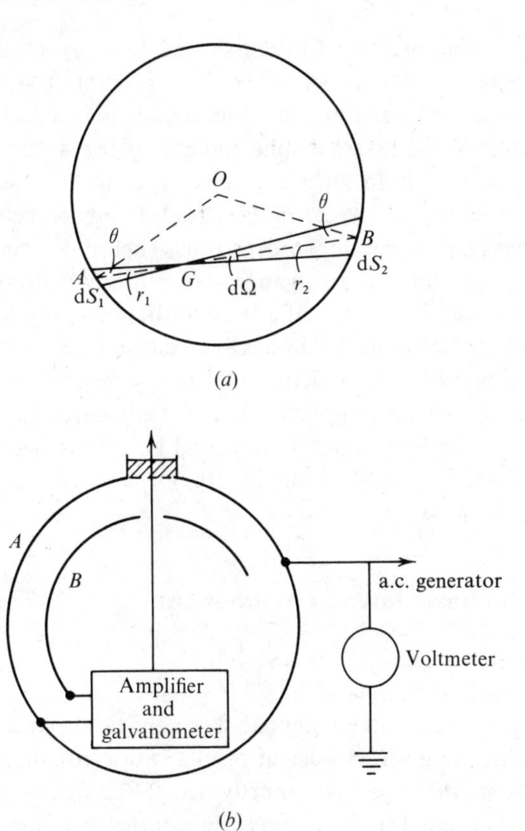


Fig. 10.1

cutting off areas $dS_1$ and $dS_2$ respectively surrounding points $A$ and $B$ on the surface of the sphere. The cone has solid angle $d\Omega$ and $A$ and $B$ are distant $r_1$ and $r_2$ respectively from the vertex $G$ of the cone. If $O$ is the centre of the sphere,

$$\angle OAB = \angle OBA = \theta$$

and

$$d\Omega = (dS_1 \cos\theta)/r_1^2 = (dS_2 \cos\theta)/r_2^2$$

Thus, if the law of force is that indicated by (10.1), the field strength at $G$ resulting from the charges on $dS_1$ and $dS_2$ will be

$$dE = \frac{Q\,d\Omega}{4\pi\epsilon_0 \cos\theta}\left[\frac{1}{r_1^{n-2}} - \frac{1}{r_2^{n-2}}\right]$$

and this will not be zero unless $n = 2$. The whole sphere can be divided by similar elementary cones and it is easy to see that their contributions to the total field at $G$ will not cancel. If, therefore, it can be shown that there is no field within the sphere, it may be concluded that $n = 2$.

This result was used by Cavendish and later by Maxwell to test the inverse-square law. More recently, a more accurate test has been carried out by Plimton and Lawton, whose method is indicated in fig. 10.1(*b*).

The charge on the external spherical conductor $A$ was varied periodically by connecting it to an a.c. generator. Any field within the sphere would give rise to an alternating potential difference between $A$ and the hemispherical conductor $B$. To detect any such p.d., $A$ and $B$ were connected to an amplifier and galvanometer and any deflection of the latter could be observed by motion of a beam of light passing through a hole in the top of the sphere. This hole was covered by a wire grid immersed in salt solution, to provide a conducting film over the whole area. Alternating voltages up to 3 kV were applied between $A$ and earth but no p.d. between $A$ and $B$ was found, although 1 $\mu$V would have been detected.

It was therefore concluded that in (10.1) $n$ cannot differ from 2 by more than one part in $10^9$.

### 10.1.3 The basic laws of magnetism

The relation

$$\boldsymbol{B} = \frac{\mu_0 I}{4\pi}\oint\frac{\mathrm{d}\boldsymbol{l}\times\boldsymbol{r}_0}{r^2} \tag{10.2}$$

has been used to calculate magnetic flux densities in innumerable experiments which have given consistent results with errors not exceeding, say, a few parts in $10^3$. We shall shortly see (§10.2.2) that the relation has been used by national standardizing laboratories in a number of countries, for the calculation of the mutual inductances of coils to be used in setting up standards of current and resistance. The agreement reached when standards produced by different laboratories (using coils of different sizes and shapes) are compared, suggests that any error in (10.2) cannot exceed a few parts in $10^6$.

In some of this work, these coils are used in alternating-current bridges, whose operation depends on the validity of Faraday's law,

$$\text{e.m.f.} = -\mathrm{d}\phi/\mathrm{d}t$$

The results thus constitute a verification of this law, with an error not exceeding a few parts in $10^6$.

### 10.1.4 Forces on currents and charged particles in a magnetic field

Here we are concerned with the two equations.

$$\boldsymbol{F} = Q\boldsymbol{u}\times\boldsymbol{B} \tag{10.3}$$

and
$$F = \oint_s I\mathrm{d}\boldsymbol{s} \times \boldsymbol{B} \qquad (10.4)$$

These equations are constantly used in the design of electrical instruments where discrepancies between theory and practice of, say, one per cent would be noticed. For more accurate measurements of the force on current-carrying conductors in a magnetic field, we may turn to current balances which have been constructed by national standardizing laboratories for the practical realization of the ampere (§10.2.2). Intercomparison of results obtained with these instruments suggests that any error in (10.4) cannot exceed about four parts in $10^6$.

Direct measurements of forces on charged particles moving through a magnetic field are more difficult to make with high accuracy. The best direct measurements of the ratio of charge to mass, for an electron, are in error by less than one part in $10^3$ and thus constitute a test of (10.3) to this accuracy. Much more accurate values of this ratio can be found by indirect means, but these are outside the scope of this book.

### 10.1.5 Maxwell's theory; the velocity of light

Maxwell's theory predicting the existence of electromagnetic waves has been abundantly verified by experiments with radio waves of all frequencies. However, we still need experimental evidence for the accuracy of the equation
$$c = 1/\sqrt{(\mu_0 \epsilon_0)} \qquad (10.5)$$
where $c$ is the velocity of light. For our purpose, two different types of measurement are relevant. In a method developed by Bergstrand, visible light from a steady source is chopped into short pulses, separated by exactly known time intervals. The light is focused to form a parallel beam, which is reflected back to the apparatus by a mirror at a known distance of the order of 10 km. Thc interval between pulses is adjusted to be exactly equal to the time taken for the light to travel to the mirror and back, and the speed of propagation can then be determined.

In a quite different method developed by Froome, microwaves with wavelengths of the order of 1 cm were generated by electrical means and their frequencies and wavelengths were precisely measured. Their speed of propagation could then be determined from the relation
$$\text{speed} = \text{frequency} \times \text{wavelength}$$

Measurements made by these two methods show excellent agreement to within one part in $10^6$, thus verifying Maxwell's predictions that light consists of electromagnetic waves and that the speed of propagation is independent of frequency. The present recommended value for the speed is

$$c = 2.997925 \times 10^8 \text{ m s}^{-1} \tag{10.6}$$

with an uncertainty which is unlikely to exceed one part in $10^6$.

The permeability of free space is, by definition (§4.3.2), exactly equal to $4\pi \times 10^{-7}$ H m$^{-1}$, so (10.5) and (10.6) give

$$\epsilon_0 = 8.854185 - 10^{-12} \text{ F m}^{-1} \tag{10.7}$$

also with an accuracy of 1 part in $10^6$.

### 10.1.6 Conclusion

In the above sections we have given a very brief outline of some of the experimental evidence which supports the basic equations of electromagnetic theory. Much additional evidence could be adduced and the precision with which measurements can be made is continually increasing. Enough has been said to show that our theory rests on an extremely sound experimental foundation.

## 10.2 Realization of the electrical units

### 10.2.1 Introduction

In this section we shall be concerned with the problems facing a national standardizing laboratory which may be called upon to calibrate an electrical instrument such as an ammeter, a voltmeter or a resistor with high precision, in terms of the appropriate SI unit. The definitions of the units have been given at various places in this book, but it will readily be appreciated that a definition such as that given for the ampere (§2.1.1) can, in practice, only be realized by indirect means, which must now be considered.

In the SI system, the units of length, mass and time are defined as follows.

The *metre* is the length equal to 1 650 763.73 wavelengths in vacuum of the radiation corresponding to the transition between the levels $2p_{10}$ and $5d_5$ of the krypton-86 atom. Distances of the order of a metre can be measured in terms of this unit with an accuracy of about one part in $10^8$.

The *kilogram* is the unit of mass; it is equal to the mass of the international prototype of the kilogram. Masses of the order of a kilogram can be compared with an accuracy of about one part in $10^9$.

The *second* is the duration of 9 192 631 770 periods of the radiation corresponding to the transition between the two hyperfine levels of the ground state of the caesium-133 atom. Times of the order of a second, and a wide range of frequencies, can be measured with an accuracy of about one part in $10^{11}$.

We shall now indicate some of the methods which have been used to establish the electrical units. This is a large subject and we shall do no more than indicate the principles of the methods, with no attempt to describe experimental details.

### 10.2.2 The ampere

The establishment of the ampere involves two quite separate problems. It is necessary to have some means of measuring, with high precision, the current flowing in a particular circuit but, since this measurement is likely to be difficult and time-consuming, it is also necessary to have some simple means of reproducing the current on other occasions. It is convenient to deal first with this second problem.

An obvious choice for the transfer instrument would be an ammeter which, once calibrated, could be used to reproduce currents at will. However, the best pointer instruments have errors of one part in $10^3$ and, even if reflecting instruments are used, the error is unlikely to be less than one in $10^4$. We need higher precision than this, so the method employed is to use a potentiometer to compare the voltage drop across a resistor of known value, through which the measured current is flowing, with the e.m.f. of a standard cell.

The ampere is defined in terms of the force between infinite parallel wires through which current is flowing (§2.1.1). In practice, to establish the ampere, we measure the force between coils whose mutual inductance can be accurately calculated, on an instrument known as a current balance. We have seen (§8.2.3) that the force is given by the relation

$$F_x = I_1 I_2 \frac{\partial M}{\partial x} \tag{10.8}$$

The essential features of a current balance are shown diagrammatically in fig. 10.2. A mutual inductor is associated with each arm of the balance. It consists of two identical coaxial fixed coils $AA$ and a movable coil $B$, coaxial with $AA$, which is attached to the arm of the balance. The same current $I$ passes through all the coils, being led to $B$, on each arm, through some hundreds of very fine silver wires attached to the balance near its fulcrum, so that they exert negligible torque. The directions of the windings are such that the torques, produced by the interactions of the currents in the fixed and movable coils respectively, reinforce each other, and the

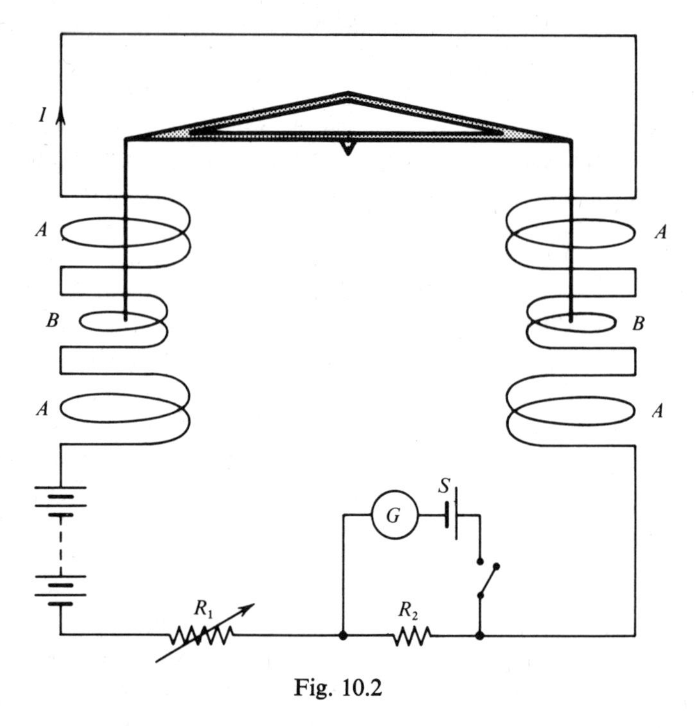


Fig. 10.2

total torque is balanced by placing weights in the scale pans. Knowing the value of the acceleration due to gravity and $\partial M/\partial x$, the current $I$ can be determined.

To enable the dimensions of the coils to be measured with the necessary accuracy, single-layer coils are used, the wire being laid in grooves cut in formers made of marble or fused silica. The mutual inductances can then be calculated from Neumann's formula (§6.7.3). For an accurate determination of current, some thousands of measurements, to a fraction of a micrometre, must be made of the diameters and axial positions of the coils, and this is likely to occupy many months.

In the final stage of the work, the current $I$ is adjusted by means of the resistor $R_1$ (fig. 10.2) until the potential drop across the resistor $R_2$, usually about one ohm, exactly balances the e.m.f. $E$ of the standard cell $S$. $I$ is then measured with the balance and we have

$$E = R_2 I \tag{10.9}$$

We shall shortly show that the value of $R_2$ can be found by other methods, so what the current balance enables us to do is to calibrate the standard cell.

The accuracy of measurement of $I$ is about four parts in $10^6$.

### 10.2.3 The absolute measurement of resistance

Without going into details, it can be stated that alternating-current bridges can be constructed which enable one to compare a resistance with a self-inductance, a mutual inductance or a capacitance, with very high precision. The comparison involves a knowledge of the frequency at which the bridge is operated but, as we have seen, frequency can be measured with more than adequate accuracy for the present purpose. In the past, both self- and mutual inductors, whose inductances can be calculated from a knowledge of their dimensions, have been used for the absolute measurement of resistance by a bridge method, with a probable error of about 2 parts in $10^6$.

Although, in principle, it is possible to construct a capacitor whose capacitance can be calculated from its dimensions, it was not known until fairly recently how this could be done with sufficient precision. A way out of the difficulty has now been found and the absolute measurement of resistance in terms of capacitance and frequency is being investigated. It is hoped that the error will be less than one part in $10^6$.

### 10.2.4 Material standards of resistance and e.m.f.

Standard resistors, of nominal value one ohm, are made of wire of an alloy such as manganin, which has both a low temperature coefficient of resistance and a low thermoelectric e.m.f. against copper. The wire is wound on formers, with as little constraint as possible, and the resistor is provided with current and potential terminals so that contact and lead resistances can be eliminated. Experience shows that, with proper precautions, such resistors will have a drift of resistance not greater than one part in $10^7$ per year. Moreover, they can be intercompared with at least this accuracy. It is the practice of a national standardizing laboratory to keep a small number of such resistors, which are compared with each other from time to time and, at less frequent intervals, have their absolute resistances measured by a bridge method.

In a rather similar manner, a group of Weston standard cells forms the working standard of e.m.f. in a national laboratory. When properly aged, the e.m.f. of a single cell will drift by less than one part in $10^6$ per year, though the e.m.fs. of different cells may differ by a few parts in $10^5$. From time to time the cells are compared with each other and, at intervals of about ten years, their e.m.fs. are measured with the current balance.

Resistors and standard cells are readily portable, so the standards of different laboratories can be compared.

## 10.3 Eddy currents

### 10.3.1 Introduction

We have seen earlier (§4.4.6) that, when any conductor is situated in a changing magnetic field, eddy currents are likely to flow in the conductor. Such currents cause a waste of power which can be of great significance in alternating-current apparatus, so it is important to know what steps can be taken to reduce them.

The calculation of eddy currents is usually difficult, but we shall now examine two simple cases where the necessary equations can be solved. The setting up of the equations provides an interesting exercise in the application of Maxwell's equations.

### 10.3.2 Eddy currents in transformer laminations

The core of a transformer carries an alternating flux density which tends to cause eddy currents to flow in the core. At low frequencies the core is normally made of an iron alloy and it is common practice to build it up from thin laminations, which are insulated from each other. This reduces the eddy currents and the consequent loss of power which they cause. With certain simplifying assumptions, it is easy to calculate the effects of the eddy currents.

In fig. 10.3 we consider a plane rectangular plate of width $h$, length $l$, thickness $2d$ and resistivity $\rho$, subjected to a magnetic field parallel to its length and varying sinusoidally with time, so that

$$B = B_0 \cos \omega t \tag{10.10}$$

We assume that $d$ is very small in comparison with $h$ and that any flux density produced by the eddy currents themselves is negligible compared with $B$. Then the eddy currents will flow in loops such as the one shown in the figure. The area enclosed by this loop is approximately $2yh$ and the resistance $R$ of the eddy-current path $2h\rho/l\,\mathrm{d}y$. Hence the e.m.f. induced in the loop is

$$E = -\mathrm{d}\phi/\mathrm{d}t = 2yh\omega B_0 \sin \omega t$$

Averaging over a cycle, the resulting mean power loss is

$$\mathrm{d}P = \tfrac{1}{2}(2yh\omega B_0)^2/R = hl\omega^2 B_0^2 y^2\, \mathrm{d}y/\rho$$

For the whole plate, the total power loss is

$$P = \frac{hl\omega^2 B_0^2}{\rho} \int_0^d y^2\, \mathrm{d}y = \frac{hl\omega^2 B_0^2 d^3}{3\rho}$$

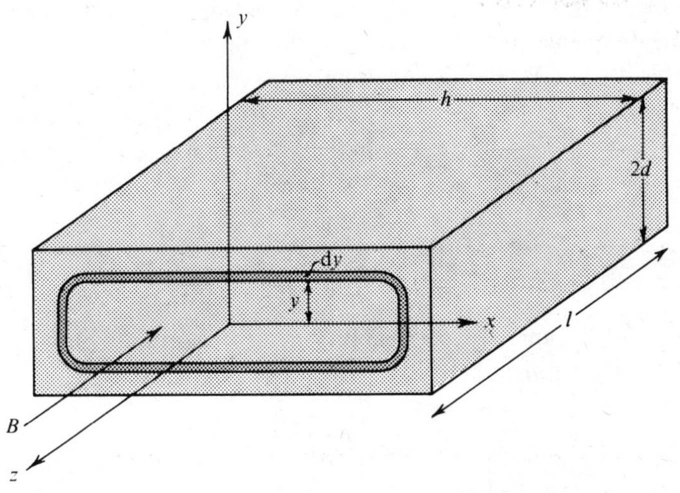


Fig. 10.3

If $f$ is the frequency, $\omega = 2\pi f$ and the volume of the plate is $2hld$. Thus we may write for the power loss per unit volume, $P_0$

$$P_0 = 2\pi^2 f^2 d^2 B_0^2/3\rho \text{ W m}^{-3} \tag{10.11}$$

Experiment shows that (10.11) gives quite accurately the functional relation between $P$, $B_0$, $f$, $d$ and $\rho$ but, because of our simplifying assumptions, the numerical constant may be in error by a factor as great as two. Although the expression has been derived for transformer sheet steel, it is equally true for a conducting sheet of any material.

Equation (10.11) shows that, for a given volume of magnetic core, eddy-current loss is directly proportional to the square of the thickness of the laminations and inversely proportional to the resistivity of the material. Very thin laminations of nickel–iron alloys can be used at frequencies up to about 10 kHz but, at still higher frequencies, ferrites with their very much higher resistivities are commonly employed.

So far we have assumed that the flux density and the magnetic field strength are constant throughout the conducting sheet but, when $d$ and/or $f$ is large enough, the eddy-currents will be sufficiently great to make an appreciable contribution to $H$ and therefore to $B$. We shall now indicate how this effect is to be taken into account.

Referring to fig. 10.3, we suppose $h$ and $l$ to be so large compared with $d$ that we are dealing with an infinite sheet, in which $H$, $B$ and the current density $J$ show no variation in the $x$- and $z$-directions, though all three will vary with $y$. $B$ and $H$ will always lie in the $z$-direction, while $J$ will always be parallel to the $x$-axis. For this one-dimensional problem, if we assume that the frequency is sufficiently low for displacement current to be negligible, two of Maxwell's equations,

$$\nabla \times \boldsymbol{E} = -\frac{\partial \boldsymbol{B}}{\partial t} \quad \text{and} \quad \nabla \times \boldsymbol{H} = \boldsymbol{J} + \frac{\partial \boldsymbol{D}}{\partial t}$$

reduce to

$$\frac{\partial E_x}{\partial y} = \frac{\partial B_z}{\partial t} \quad \text{and} \quad \frac{\partial H_z}{\partial y} = J_x \tag{10.12}$$

Also

$$E_x = \rho J_x \quad \text{and} \quad B_z = \mu H_z \tag{10.13}$$

and we shall assume the permeability $\mu$ to be constant. Eliminating $J_x$ and $B_z$ from (10.12) and (10.13) we then have

$$\frac{\partial^2 H_z}{\partial y^2} = \frac{\mu}{\rho}\frac{\partial H_z}{\partial t} \tag{10.14}$$

We are concerned with the case where the external field applied to the sheet varies sinusoidally with time and is represented by

$$H = H_0 \cos \omega t \tag{10.15}$$

The required boundary conditions are therefore

$$H_z = H_0 \cos \omega t \quad \text{when} \quad y = \pm d$$

The solution of (10.14) to satisfy these conditions is straightforward, but tedious, and will not be reproduced here. If we put

$$\sqrt{(\mu\omega/2\rho)} = \beta \tag{10.16}$$

the end result is

$$H_z = H_0 \left(\frac{\cosh 2\beta y + \cos 2\beta y}{\cosh 2\beta d + \cos 2\beta d}\right)^{\frac{1}{2}} \cos(\omega t - \theta)$$

where

$$\tan \theta = \frac{\sinh \beta(d-y) \sin \beta(d+y) + \sinh \beta(d+y) \sin \beta(d-y)}{\cosh \beta(d-y) \cos \beta(d+y) + \cosh \beta(d+y) \cos \beta(d-y)} \tag{10.17}$$

At the centre of the sheet

$$H_{zc} = \frac{2^{\frac{1}{2}} H_0 \cos(\omega t - \theta_c)}{(\cosh 2\beta d + \cos 2\beta d)^{\frac{1}{2}}}$$

$$\tan \theta_c = \tanh \beta d \tan \beta d \tag{10.18}$$

As an example of the application of (10.18) we take the following values for the silicon–iron alloy commonly used in power transformers

$$\rho = 6 \times 10^{-7}\,\Omega\text{m}, \quad \mu_r = 3000, \quad d = 2.5 \times 10^{-4}\,\text{m}$$

giving

$$\mu = \mu_0 \mu_r = 3.77 \times 10^{-3}$$

At a frequency of 50 Hz, $H_{zc}$ is almost exactly the same as $H_0$ but, if we raise the frequency to 5 kHz, we find that $H_{zc}$ has fallen to 0.17 $H_0$. The simple calculation of core loss from (10.11) would clearly be inapplicable in this second case. A more satisfactory calculation could be carried out, using (10.12) and (10.17) to find the variation of $J_x$ with $y$, but even this would not be accurate, since we have assumed that $\mu$ is constant. Perhaps the greatest value of the foregoing theory is to indicate the maximum frequency at which any particular lamination is likely to be satisfactory.

### 10.3.3 Eddy currents in wires through which alternating current is flowing

When alternating current flows in a long cylindrical wire of resistivity $\rho$, it produces an alternating magnetic field with circular flux lines whose centres lie on the axis of the wire. This field generates eddy currents parallel to the axis and, since the total eddy current, integrated over a cross-section, must be zero, the direction of eddy-current flow must reverse at some particular radius. Elementary considerations show that the eddy current opposes the main current along the axis and reinforces it near the surface of the wire. As a result, the total current density increases from the axis outwards and, at sufficiently high frequencies, the current is almost completely concentrated in a thin layer near the surface of the wire. This concentration is referred to as the *skin effect*; it causes a considerable increase in the effective resistance of the wire and an almost complete removal of its internal self-inductance.

To set up an equation for the variation of current density with radius, it will be appropriate to work with cylindrical coordinates and it is left as an exercise for the reader to show that, in this system, with unit vectors $\boldsymbol{a}_r$, $\boldsymbol{a}_\theta$ and $\boldsymbol{a}_z$, the curl of any vector $\boldsymbol{A}$ becomes

$$\nabla \times \boldsymbol{A} = \boldsymbol{a}_r\left(\frac{1}{r}\frac{\partial A_z}{\partial \theta} - \frac{\partial A_\theta}{\partial z}\right) + \boldsymbol{a}_\theta\left(\frac{\partial A_r}{\partial z} - \frac{\partial A_z}{\partial r}\right) + \boldsymbol{a}_z\left(\frac{\partial A_\theta}{\partial r} + \frac{A_\theta}{r} - \frac{1}{r}\frac{\partial A_r}{\partial \theta}\right) \tag{10.19}$$

In our system there is no variation with respect to $\theta$ or to $z$, the only component of $\boldsymbol{B}$ or of $\boldsymbol{H}$ is $B_\theta$ or $H_\theta$, and the only component of $\boldsymbol{E}$ or $\boldsymbol{J}$ is $E_z$ or $J_z$. Thus, assuming displacement current to be negligible, the appropriate Maxwell equations

$$\nabla \times \boldsymbol{E} = -\frac{\partial \boldsymbol{B}}{\partial t} \quad \text{and} \quad \nabla \times \boldsymbol{H} = \boldsymbol{J}$$

become

$$\frac{\partial E_z}{\partial r} = \frac{\partial B_\theta}{\partial t} \quad \text{and} \quad \frac{\partial H_\theta}{\partial r} + \frac{H_\theta}{r} = J_z \tag{10.20}$$

with $$B_\theta = \mu H_\theta \quad \text{and} \quad E_z = \rho J_z \tag{10.21}$$

Eliminating $B_\theta$, $H_\theta$ and $E_z$ from these equations gives

$$\frac{\partial^2 J_z}{\partial r^2} + \frac{1}{r}\frac{\partial J_z}{\partial r} - \frac{\mu}{\rho} J_z = 0 \tag{10.22}$$

The solution of this equation in terms of Bessel functions is quite difficult and will not be reproduced here.† It enables one to calculate the variation of the resistance of a straight wire with frequency and shows that the skin effect is far from negligible. Thus for copper wire, of one millimetre diameter, the resistance at a frequency of 1 MHz would be about four times the direct-current value.

Although calculations based on (10.22) are valid for a long straight wire, such as the inner conductor of a concentric cable, they have little relevance to wire in a coil, where the magnetic field acting on any element depends not only on the current in that element, but also on the current in all adjacent elements. This *proximity effect*, which is usually difficult to calculate, causes an additional variation of resistance with frequency.

Since, at a given frequency, both skin effect and proximity effect increase rapidly with wire diameter, they can both be reduced by using stranded wire, with each strand insulated from the others. This technique is often used when it is desired to construct inductors of low resistance to carry current of high frequency.

## 10.4 The behaviour of practical inductors, capacitors and resistors in alternating-current circuits

### 10.4.1 Introduction

Hitherto we have spoken of resistance, capacitance and inductance as though each of these was the unique property of an electrical component termed a resistor, a capacitor or an inductor, as the case might be. In fact, any component that we can construct in practice possesses all three of the above properties and so does not behave as an ideal component would be expected to do. This can be of considerable importance when the component is used in alternating-current circuits and we shall now indicate very briefly the behaviour to be expected from real components.

† See, for example, N. W. McLachlan, *Bessel functions for engineers* (Oxford, Clarendon Press, 1955).

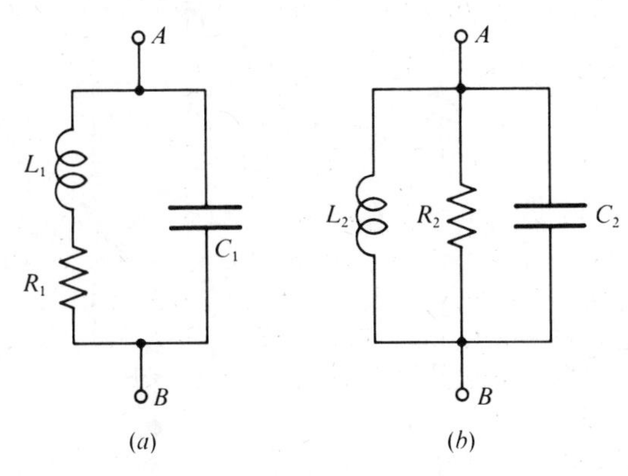


Fig. 10.4

### 10.4.2 Real inductors

In an a.c. circuit, the function of a perfect inductor is to provide a reactance in which the current lags the voltage by 90°. A practical inductor necessarily has resistance also and one measure of its perfection is its *quality factor* $Q$ where

$$Q = \frac{\text{reactance}}{\text{series resistance}} = \frac{L\omega}{R} \tag{10.23}$$

Unless we are concerned with very low frequencies (below 50 Hz, say), it is usual for $Q$ to be greater than 10 and values of several hundreds are common.

An inductor often has a core of magnetic material and, if the inductance is to be nearly independent of current, it is necessary to have an air gap in the magnetic circuit, so that the total reluctance is largely determined by the gap and is unaffected by the varying permeability of the material.

Capacitance must exist between each element of length of the winding and all other elements, with the overall result that a voltage applied to the terminals of the inductor causes a current, leading the voltage by 90°, to flow through the stray capacitance.

One can devise several equivalent circuits to take account of these various effects. Two are shown in fig. 10.4; which is the more convenient depends on the circuit of which the inductor forms a part. The total impedance between $A$ and $B$ must be the same for the two circuits and it is easy to show that, if $Q$ exceeds 10, one may write with an error less than one per cent,

$$L_1 = L_2 = L, \quad C_1 = C_2, \quad R_2 = L^2\omega^2/R_1 = Q^2R_1 \tag{10.24}$$

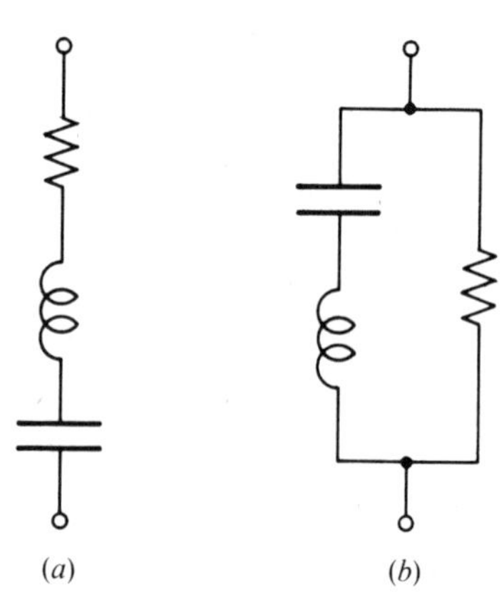

(*a*) (*b*)

Fig. 10.5

The resistances $R_1$ and $R_2$ include contributions from the direct-current resistance of the wire, eddy currents in the wire and core, magnetic hysteresis in the core and, possibly, dielectric hysteresis in the material through which the capacitive current flows. All of these cause power loss and so can be lumped together. Strictly, the hysteresis losses are non-linear and should be represented by a non-linear contribution to $R_1$ and $R_2$, but it is rarely necessary to include this refinement.

$L_1$, $L_2$, $C_1$ and $C_2$ generally remain sensibly constant over a wide range of frequencies, but eddy-current and hysteresis losses increase with frequency, so $R_1$ will increase and $R_2$ will decrease as the frequency rises. It thus comes about that the $Q$-factor often does not vary very rapidly with frequency.

At some frequency $f_0$, resonance will occur between $L_1$ and $C_1$ (or $L_2$ and $C_2$) and, above this frequency, the component will behave as a capacitor rather than an inductor. For a coil of a few thousand turns, wound on a core of nickel–iron alloy, $f_0$ may not exceed a few kilohertz.

### 10.4.3 Real capacitors

Rather similar considerations apply to capacitors, where the dielectric may not have infinite resistivity and will have some hysteresis loss. The resistance of the conductors may not be negligible and, particularly if they are in the form of a roll, inductance may be appreciable. Suitable equivalent circuits are shown in fig. 10.5. For most purposes the inductance can be ignored though, if it were required to provide a low reactance over a wide range of frequencies, it might be profitable to connect, for example, an electrolytic capacitor (with large capacitance, but also appreciable inductance) in parallel with a mica capacitor (with small capacitance and negligible inductance).

The losses in a capacitor cause the phase difference between current and voltage to fall short of $\pi/2$ by some small *loss angle* $\delta$. The value of $\delta$, expressed in radians, is taken as a measure of the imperfection of the capacitor. Except for electrolytic capacitors, it is usually less than 0.01, and often very much less. Thus, with sufficient accuracy,

$$\delta = \underset{\text{(fig. 10.5(a))}}{\frac{\text{series resistance}}{\text{reactance}}} = \underset{\text{(fig. 10.5(b))}}{\frac{\text{reactance}}{\text{parallel resistance}}}$$

### 10.4.4 Real resistors

With resistors, we have to consider two quite different situations. In many electronic circuits the precise value of a resistor is of little importance and the resistors consist of extremely thin films of metal or oxide, or are made of some composite material of very high resistivity. In such resistors inductance is rarely of importance. There is, however, stray capacitance in parallel with the resistor and this may become significant at very high frequencies or when the value of resistance is very high.

A quite different problem arises in precision resistance boxes, where inductance and capacitance should be negligible and resistance values should remain constant from zero frequency to at least 10 kHz. The resistors are made with wire such as manganin or constantan, which has high resistivity, and it is not difficult to choose wire diameters such that skin effect and proximity effect are negligible. Winding the wire on thin plane formers reduces the inductance but does not eliminate it.

In an early attempt to reduce the inductance still further, the wire was bent back on itself (fig. 10.6(*a*)), so that 'go' and 'return' currents were always close together and thus produced little magnetic field. However, this design greatly increases the capacitance between the ends of the resistor and is suitable only for resistance values of a few ohms. For values up to a few hundred ohms, the scheme shown in fig. 10.6(*b*) is satisfactory. In this, two single-layer resistors wound in opposite directions are connected in parallel, so that the magnetic field of each is largely cancelled by that of the other. Moreover, adjacent elements of the two windings are always at the same potential, so capacitance effects are very small.

For still higher values, a multi-layer single coil is used. This has both inductance and capacitance, with an equivalent circuit as shown in fig. 10.6(*c*). It is easy to show that, if

$$R \gg L\omega \quad \text{and} \quad L = RC \tag{10.25}$$

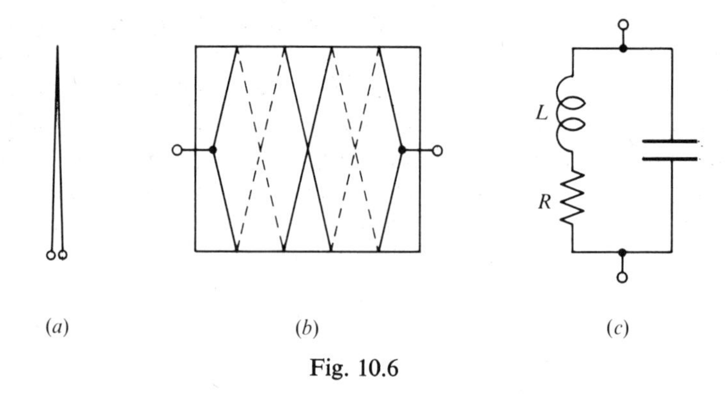


Fig. 10.6

the inductance and capacitance effects cancel each other almost completely, so that the total impedance is a nearly pure resistance $R$. It is found that the conditions of (10.25) can be satisfied by proper choice of wire gauge, thickness of former and disposition of turns.

Using techniques such as these, resistance boxes can be constructed which are satisfactory at frequencies up to about 50 kHz.

## 10.5 The non-relativistic motion of a charged particle in electric and magnetic fields

### 10.5.1 Introduction

When a charged particle is moving through an electrostatic or a magnetic field it experiences a force which, in principle, can be calculated by the methods discussed earlier. In general, this force will vary from point to point and the calculation of the trajectory of the particle may be a matter of great difficulty. In this section we discuss a few simple cases which can be treated analytically and describe qualitatively some others which arise in practice.

These matters do not form part of what is normally understood by electromagnetic theory. They are included here partly because of their intrinsic importance and partly because they provide examples of the application of electromagnetic theory.

### 10.5.2 Energy in an electrostatic field

If a particle of mass $m$ has charge $Q$ and is situated in an electrostatic field of strength $\boldsymbol{E}$, it experiences a force $\boldsymbol{F}$, where

$$\boldsymbol{F} = Q\boldsymbol{E} \tag{10.26}$$

If the particle is not otherwise constrained, it will move under the influence of this force and will gain kinetic energy, which may be supplied either by a change in the energy stored in the field itself or, if the field is kept constant by means of batteries connected to electrodes, by transient currents flowing through these batteries. If the particle is initially at rest and moves from a point where the potential is $V_1$ to a second point where the potential has some lower value $V_2$, we may equate its loss of potential energy to its gain of kinetic energy and write

$$Q(V_1 - V_2) = \tfrac{1}{2}mv^2 \qquad (10.27)$$

where $v$ is the speed of the particle at the second point.

This equation depends on Newtonian mechanics and is valid only if $v$ is very small in comparison with the speed of light. When this condition is not fulfilled, relativistic mechanics, taking account of the variation of the mass of a particle with its speed, must be used. In all that follows it will be assumed that relativistic effects can be neglected.

If in (10.27) we insert the values of $Q$ and $m$ for an electron, we find

$$\begin{aligned} v &= \surd[2Q(V_1 - V_2)/m] \\ &= 5.93 \times 10^5 \surd(V_1 - V_2) \text{ m s}^{-1} \end{aligned} \qquad (10.28)$$

Thus, for an electron, relativistic effects can be ignored if $(V_1 - V_2)$ is less than, say, 5 kV and we shall consider only those situations for which this is the case.

A convenient unit for specifying the kinetic energy of an electron is the *electron volt*. It is the gain in kinetic energy which an electron experiences when accelerated through a potential difference of one volt. It is equal to an energy of $1.602 \times 10^{-19}$ joule.

### 10.5.3 The electron gun

In many pieces of electrical equipment, such as cathode-ray tubes and electron microscopes, we need an electron beam of narrow angle, in which all of the electrons are moving with very nearly the same velocity. A device for producing such a beam is known as an *electron gun* and one type of gun, which is widely used, is shown in fig. 10.7.

Electrons are emitted from a thermionic *cathode* $C$ which might, for example, be a directly heated loop of tungsten wire or an indirectly heated oxide-coated cathode. Surrounding $C$ is an electrode $G$, known as the *grid*, which has cylindrical symmetry about the axis of the system and is usually placed so that the tip of $C$ lies just inside $G$. In front of $G$ lies an anode $A$ which is maintained at a positive potential $V_A$ with respect to $C$.

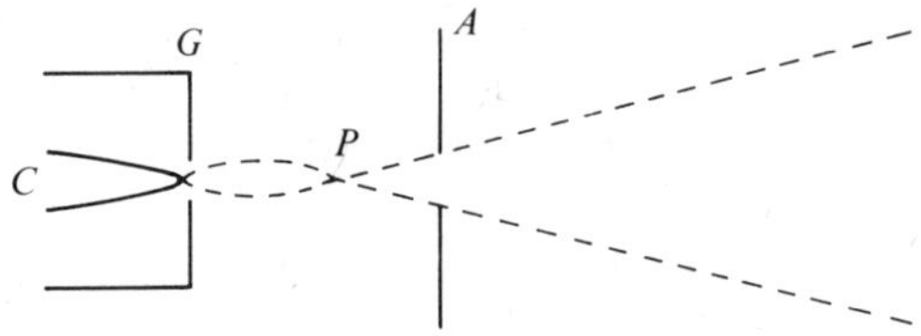

Fig. 10.7

If electrons were emitted from $C$ with zero velocity, the final velocity of those passing through the hole in $A$ would be determined by $V_A$, in accordance with (10.28). In fact, the electrons are emitted with a Maxwellian distribution of velocities, with a mean energy of one or two tenths of an electron volt, depending on the temperature of the emitter. This spread of velocities is present in the electrons which pass through $A$. If $V_A$ is of the order of 1 kV or more the spread is often of no importance, though this is by no means always the case.

The grid $G$ is biassed negatively with respect to $C$ and the extent of the bias controls the number of electrons leaving $C$ and therefore the number finally passing through $A$. The dashed lines in fig. 10.7 indicate possible electron trajectories and, from these it would appear that electrons passing through $A$ form a pencil diverging from a small area at $P$, known as the *crossover*. In some guns the crossover may be virtual; that is, the electron pencil passing through $A$ may appear to diverge from a point behind the cathode.

One practical point is worthy of mention. In the foregoing discussion the potentials of the anode and grid have been stated with reference to the potential of the cathode. If, as is generally the case, we wish to work with the beam of electrons which has passed through the anode aperture, it is usually convenient to maintain the anode at earth potential, with the cathode and grid at appropriate negative potentials with respect to earth. This is common practice, for example, in cathode-ray oscilloscopes.

There are other types of electron gun and means are available for producing narrow beams of charged particles other than electrons. We now consider how beams of this kind are affected when they enter electric or magnetic fields.

### 10.5.4 Motion in a uniform electrostatic field

We suppose the field to be established between two large parallel plane conductors, of which $A$ (fig. 10.8($a$)) is maintained at zero potential while $B$ has potential $V$. The simplest case is that in which the narrow beam of

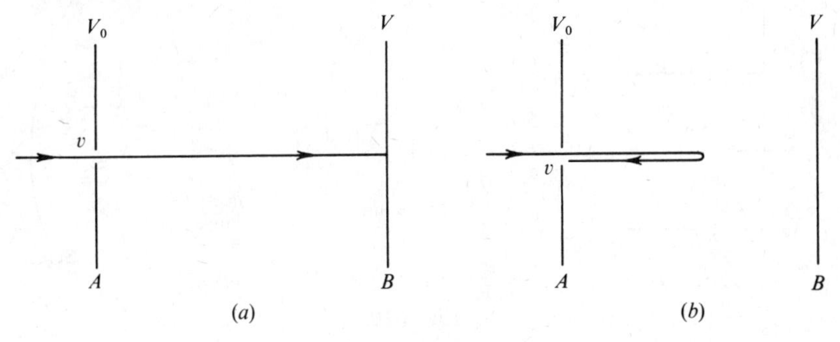


Fig. 10.8

particles, of charge $Q$ and mass $m$, enters the field through a small hole in $A$, in a direction parallel to the field. At the point of entry let the particles have velocity $v$ corresponding to acceleration through a potential difference $V_0$, so that from (10.27)

$$\tfrac{1}{2}mv^2 = V_0 Q \tag{10.29}$$

If the direction of the field between the plates is such as to cause further acceleration of the particles, these will travel without deflection until they strike $B$ with velocity $v_1$ given by

$$\tfrac{1}{2}mv_1^2 = Q(V_0+V) \tag{10.30}$$

If, on the other hand, $V$ is of opposite sign so that the particles suffer retardation between the plates, they will either strike $B$ with reduced velocity $v_2$ given by

$$\tfrac{1}{2}mv_2^2 = Q(V_0-V) \tag{10.31}$$

or, if $|V| > |V_0|$, they will be turned back before reaching $B$ (fig. 10.8(*b*)).

As a second simple case we suppose the particles, travelling with initial velocity $v$ parallel to the $x$-axis, to enter a uniform electrostatic field $E$ parallel to the $y$-axis (fig. 10.9(*a*)). The particles will experience no force in the $x$-direction, so the component of velocity in this direction will have the constant value $v$ and, at any time $t$ after entering the field, the $x$-coordinate will be $vt$. In the $y$-direction the initial velocity component is zero, but the particle has acceleration equal to $QE/m$. Its $y$-coordinates after time $t$ will therefore be $\tfrac{1}{2}QEt^2/m$. Eliminating $t$, we find for the trajectory of the particle the parabola whose equation is

$$y = QEx^2/2mv^2 \tag{10.32}$$

A situation similar to this occurs in a cathode-ray tube with electrostatic deflection (fig. 10.9(*b*)). However, calculations based on (10.32) are of limited value since considerable fringing of the field occurs at the edges of the deflector plates.

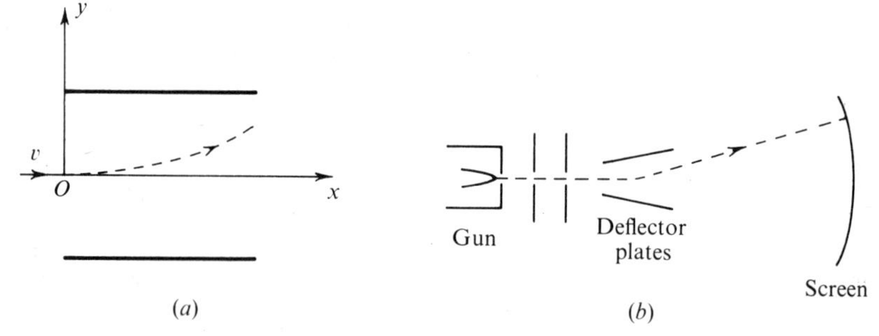


Fig. 10.9

### 10.5.5 Motion in a uniform magnetic field

We have seen earlier (§4.3.1) that the force $\boldsymbol{F}$ acting on a particle with charge $Q$ moving with velocity $v$ in a magnetic field of flux density $\boldsymbol{B}$, is given by

$$\boldsymbol{F} = Q\boldsymbol{v} \times \boldsymbol{B} \tag{10.33}$$

Thus, if the particle is travelling parallel to the field, it experiences no force. A second important deduction is that, whatever the direction of $v$, any force acting on the particle will always be perpendicular to $v$. Hence this force cannot alter the energy of the particle (or its speed), though it can change the direction in which the particle moves.

We consider the case of a particle moving with velocity $v$ at right angles to a field whose flux density has magnitude $B$ (fig. 10.10). The initial velocity $v$ is in the $x$-direction and $B$ is perpendicular to the plane of the diagram and is directed into the paper. The force $F$ is then in the $y$-direction initially, but subsequently remains constant in value and changes direction so that it is always normal to the velocity of the particle. The particle therefore moves in a circular orbit of radius $r$ such that

$$mv^2/r = QvB$$

or

$$r = mv/QB \tag{10.34}$$

In a cathode-ray tube employing magnetic deflection, coils placed on the outside of the tube produce a roughly constant magnetic field over a limited region of space and the electron beam is deflected as indicated in fig. 10.11. Once again, calculations of deflection based on (10.34) are rendered inaccurate by fringing of the field.

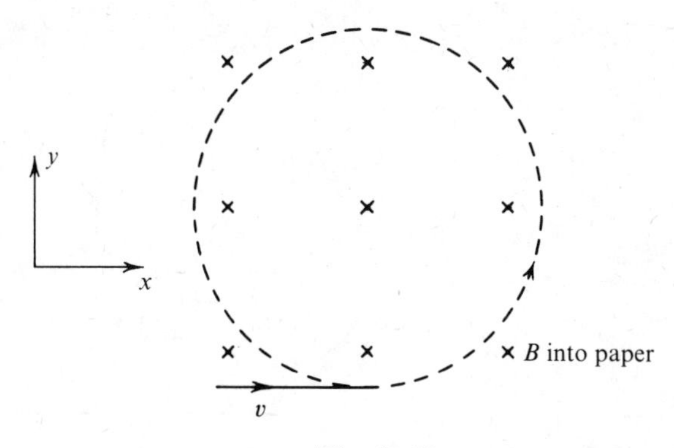


Fig. 10.10

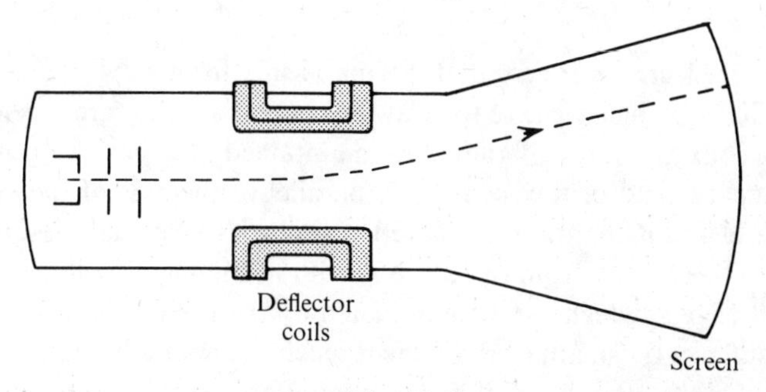


Fig. 10.11

### 10.5.6 Motion in combined electric and magnetic fields

When a charged particle moves in a combined electric and magnetic field, we may formally write down its equations of motion in rectangular coordinates as

$$\begin{aligned}\frac{d^2x}{dt^2} &= \frac{Q}{m}\left(E_x + B_z\frac{dy}{dt} - B_y\frac{dz}{dt}\right)\\ \frac{d^2y}{dt^2} &= \frac{Q}{m}\left(E_y + B_x\frac{dz}{dt} - B_z\frac{dx}{dt}\right)\\ \frac{d^2z}{dt^2} &= \frac{Q}{m}\left(E_z + B_y\frac{dx}{dt} - B_x\frac{dy}{dt}\right)\end{aligned} \tag{10.35}$$

Except in the simplest cases, a computer program is generally needed to find the field components and to solve the resulting equations.

It is sometimes convenient to employ other systems of coordinates and, as an example of a two-dimensional problem that can be partially solved, we consider the cylindrical magnetron, shown in section in fig. 10.12(*a*).

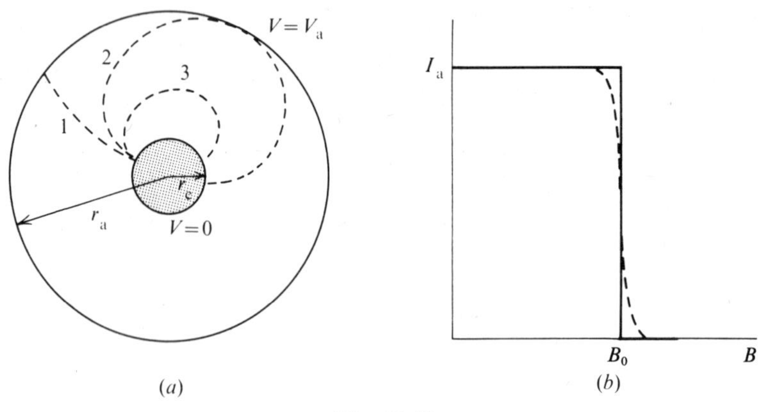


Fig. 10.12

In this device electrons are emitted thermionically from a cylindrical cathode, of radius $r_c$, which we take to be at zero potential. They are accelerated towards a coaxial anode, of radius $r_a$, maintained at a positive potential $V_a$. A magnetic field of flux density $B$, parallel to the axis of the system, causes the electrons to traverse curved trajectories. We shall assume the electrons to be emitted from the cathode with negligible velocity and shall ignore end effects, which can be eliminated by the use of guard rings.

If $B$ is sufficiently small, electrons will reach the anode by paths similar to curve 1. On the other hand, if $B$ is large enough, they will be turned back to the cathode before reaching the anode (curve 3). For a particular value $B_0$ a critical situation occurs (curve 2), where the electrons just graze the surface of the anode. Thus a plot of anode current $I_a$ against $B$ should yield the rectangular curve shown by the full line in fig. 10.12(*b*). In practice the cut-off of $I_a$ is not abrupt and the dashed curve is obtained because of the small spread of velocities with which the electrons are emitted from the cathode.

A determination of the complete trajectories is difficult, but we can readily calculate $B_0$ as follows. Using polar coordinates, the velocity $v$ of the electron is given by

$$v^2 = \left(\frac{dr}{dt}\right)^2 + \left(r\frac{d\theta}{dt}\right)^2 \tag{10.36}$$

For the critical condition (curve 2) we have

$$\frac{dr}{dt} = 0 \quad \text{when} \quad r = r_a$$

and, from energy considerations,

$$v^2 = 2QV/m \quad \text{when} \quad r = r_a$$

Hence, from (10.36),

$$\frac{d\theta}{dt} = \frac{1}{r_a}\sqrt{\frac{2QV}{m}} \quad \text{where} \quad r = r_a \tag{10.37}$$

Equating the rate of change of angular momentum of an electron to the torque acting on it, we find

$$\frac{d}{dt}\left(mr^2\frac{d\theta}{dt}\right) = rBQ\frac{dr}{dt}$$

which, on integration, gives

$$mr^2\frac{d\theta}{dt} = \tfrac{1}{2}r^2BQ + C$$

At the instant of emission from the cathode,

$$r = r_c \quad \text{and} \quad \frac{d\theta}{dt} = 0$$

so

$$C = -\tfrac{1}{2}r_c^2BQ$$

and

$$mr^2\frac{d\theta}{dt} = \tfrac{1}{2}BQ(r^2 - r_c^2)$$

For the critical field $B_0$, $d\theta/dt$ is given by (10.37), when $r = r_a$, so that finally

$$B_0 = \frac{r_a}{(r_a^2 - r_c^2)}\sqrt{\frac{8mV}{Q}} \tag{10.38}$$

Substituting the numerical values of $m$ and $Q$ for an electron,

$$B_0 = (6.74 \times 10^{-6}\, r_a\sqrt{V})/(r_a^2 - r_c^2) \text{ tesla} \tag{10.39}$$

### 10.5.7 The focusing of electron and ion beams

Hitherto we have assumed that charged particles entering an electric or magnetic field are, at the outset, all travelling in the same direction. In practice, this rarely happens: particles entering the field through a small hole (e.g. the anode of an electron gun) usually form a divergent pencil, while if they enter through a narrow slit, the beam is wedge shaped. The angle of the pencil or wedge may be large or small, depending on circumstances.

In many instruments it is desirable to cause the particles to come to a point or line focus at a place distant from the point of entry and several arrangements of electric and/or magnetic fields have been devised to achieve this end. We shall consider two of the simplest of these.

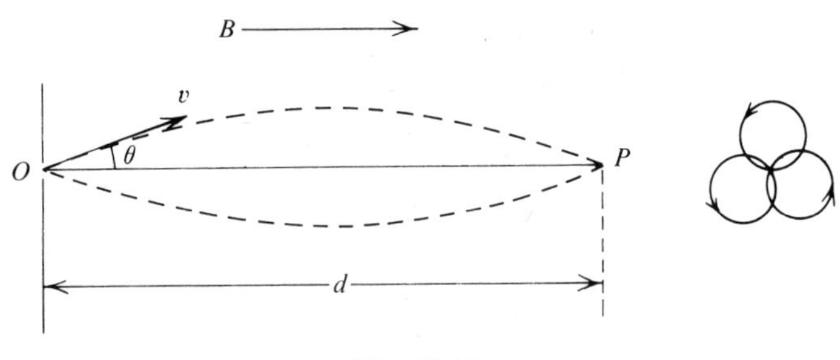


Fig. 10.13

In fig. 10.13 suppose a divergent pencil of particles to pass through a small hole $O$, to enter a uniform magnetic field of flux density $\boldsymbol{B}$ parallel to the axis of the pencil. Let $v$ be speed of the particles. We consider a particle whose initial direction makes angle $\theta$ with the axis $OP$. It will have components of velocity $v \cos \theta$ along the axis and $v \sin \theta$ at right angles to this direction. The former will be unaffected by the field while, as a result of the latter, the projection of the trajectory on a plane normal to the axis will be a circle and, from (10.34), its radius will be

$$r = mv \sin \theta / QB$$

The time taken for the particle to pass once round this circle is $2\pi m/QB$ and, during this time, the particle has moved parallel to the axis a distance

$$d = 2\pi mv \cos \theta / QB \tag{10.40}$$

The trajectory of the particle is a spiral of pitch $d$ and, after completing one turn of the spiral, the particle which set out from the axis at $O$, is again on the axis at $P$.

If the divergence of the pencil at $O$ is limited, by suitably placed apertures, so that $\theta$ never exceeds a few degrees, we can put $\cos \theta$ equal to unity and $d$ then becomes independent of $\theta$. All of the particles diverging from a point at $O$ will be brought to a focus at $P$. Other foci, for which the particles have completed 2, 3, ..., etc. turns of their spirals, will occur at points, spaced distance $d$ apart, along the axis.

For our second example we consider a wedge-shaped beam of particles entering the space between two coaxial cylinders through a narrow slit $O$ which is parallel to the axis (fig. 10.14). The initial tangential velocity of the particles is $v_0$ and a potential difference is applied between the cylinders to produce an electric field in a direction to oppose the centrifugal force on the particles as they move round the annular space. The entry slit is at a distance $r_0$ from the axis, end effects are to be neglected and the angular spread of the wedge-shaped beam is assumed to be only a few degrees.

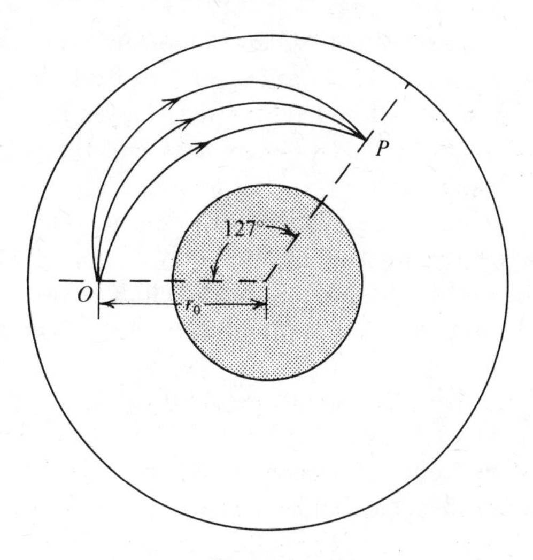


Fig. 10.14

We know (§3.6.2) that the electric field between the cylinders is inversely proportional to the radial distance from the axis. Let us adjust the potential on the cylinders so that a particle passing through the slit with zero radial velocity has centrifugal force exactly balanced by the field $E_{r_0}$, and so continues in a circular path of radius $r_0$. The condition for this is

$$E_{r_0} Q = mv_0^2/r_0 \tag{10.41}$$

and, for the field at any other radius, we then have

$$E_r = r_0 E_{r_0}/r = mv_0^2/Qr \tag{10.42}$$

We now wish to investigate the subsequent motion of particles which pass through the slit with small radial velocities in addition to their tangential velocity $v_0$. The force which the field exerts on them is entirely radial, so their initial angular momentum $mr_0 v_0$ will remain unchanged. Thus, if a particle is at distance $r$ from the axis, we have

$$mr^2 \frac{d\theta}{dt} = mr_0 v_0 \tag{10.43}$$

For the radial motion of a particle

$$m \frac{d^2r}{dt^2} = mr \left(\frac{d\theta}{dt}\right)^2 - QE_r \tag{10.44}$$

and substituting from (10.42) and (10.43) we have

$$\frac{d^2r}{dt^2} = \frac{r_0^2 v_0^2}{r^3} - \frac{v_0^2}{r} \tag{10.45}$$

The first term on the right-hand side results from the centrifugal force, while the second expresses the inward pull of the field. When $r$ is equal to $r_0$, the two terms are equal and the particle moves in a stable circular orbit. If $r$ exceeds $r_0$, the field predominates and the particle is pulled inwards. The reverse happens if $r$ is less than $r_0$.

To gain further information about the way in which this corrective action operates, we assume that $r$ is never very different from $r_0$ and write $r = r_0 + \delta r$. This will be a good approximation for a wedge-shaped beam of small angle. Substituting in (10.45) and making the usual approximations we get

$$\frac{\mathrm{d}^2(\delta r)}{\mathrm{d}t^2} + \frac{2v_0^2\,\delta r}{r_0^2} = 0$$

If we measure time from the instant which a particle passes through the slit, the solution of this equation is

$$\delta r = A \sin (2^{\frac{1}{2}} v_0/r_0)\, t \tag{10.46}$$

where the constant $A$ depends on the direction in which the particle was moving when it passed through the slit.

For all of the particles, $\delta r = 0$ when $t = 0$ and $\delta r$ will be equal to zero again when

$$(2^{\frac{1}{2}} v_0/r_0)\, t = \pi \quad \text{or} \quad t = \pi r_0/2^{\frac{1}{2}} v_0$$

In this interval the trajectory will subtend an angle $\theta$, where

$$\theta = \pi/2^{\frac{1}{2}} \text{ radians} = 127^\circ$$

Thus the particles are brought to a line focus at $P$ in fig. 10.14.

### 10.5.8 Electron lenses

In many instruments employing electron beams it is necessary for electrons diverging from the crossover of an electron gun to be brought to a focus at some other point. This requirement arises, for example in cathode-ray tubes and electron microscopes, in electron-probe instruments such as X-ray microanalysers and in television camera tubes. To achieve this result, devices known as *electron lenses* are used. In the following brief account, we shall confine our description to systems which have cylindrical symmetry about an axis, though two-dimensional systems, to bring electrons from a line source to a line focus, are sometimes needed.

If an electron lens is to be based on deflections in an electrostatic field, the field must be established by applying appropriate potentials to electrodes of suitable shape. The electrodes must have symmetry about the axis of the system and, since it must be possible for electrons to travel

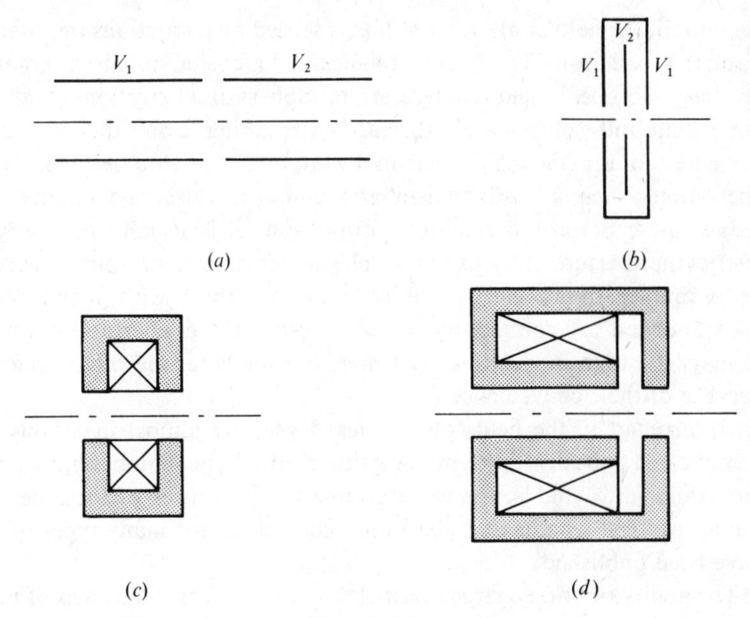


Fig. 10.15

along the axis, the shapes most commonly used are hollow cylinders and plates with circular holes. Two examples are shown in fig. 10.15. In the *two-cylinder lens* (*a*), the velocity of the electrons will be changed by passage through the lens and this is sometimes inconvenient. In such cases the *unipotential lens* (*b*) might be used.

We have already seen that an electron beam can be focused by a uniform magnetic field but, in this case, both the source and the focus are situated in the field. This is often undesirable and most magnetic lenses produce fields confined more or less completely to definite regions of space. Examples of such lenses are shown in fig. 10.15(*c*) and (*d*).

So far we have referred only to the focusing of electrons diverging from a point on the axis of a lens, but it can be shown that pencil beams from points off the axis are also brought to corresponding foci, so that an extended source produces an extended image. With electrostatic lenses the image has the same orientation as the source, but magnetic lenses produce a rotation of the image.

It is hardly to be expected that fields, whether electrostatic or magnetic, set up in the manner described above, would have the configuration needed to produce distortionless images and, in fact, aberrations in electron lenses are much more severe than in their optical counterparts. We have previously seen (§6.3.6) that, once the field on the axis of the system has been fixed, the field at all other points is settled by restrictions imposed by Laplace's equation. Thus, the designer is hampered in his attempt to produce the ideal field configuration. Spherical aberration is always important and 'chromatic' aberration, resulting from the spread of velocities in the incident electron beam, may be troublesome. Other aberrations such as astigmatism and coma are also encountered. To reduce these defects to tolerable proportions it is usually necessary to restrict the aperture of a lens to a much smaller value than would otherwise be desirable. In an electron microscope apertures with diameters of 0.1 mm or less are commonly used. As a general rule, aberrations are less in magnetic than in electrostatic lenses, but the latter are often preferred because of their convenience.

Unfortunately, the fields of practical lenses are almost invariably too complicated to be dealt with by analytical methods, but computer programs are available to enable one to determine the properties of these devices. Tables of focal lengths and aberration coefficients for many types of lens have been published.

The ability to bring charged particles to a focus forms the basis of many important instruments. Electron microscopes and electron-probe instruments have already been mentioned. In mass spectrometers, ions with a particular ratio of charge to mass are focused at one point, while those with different ratios converge to different points. Thus, the different ions can be separated and identified. Again, in electron spectrometers, electrons moving with different speeds are brought to different foci, so the distribution of velocities in the initial beam can be determined.

## 10.6 Worked examples

1. Show that current density $J$ in a space charge limited diode is given by

$$J = \frac{4\epsilon_0}{9\pi}\left(\frac{2e}{m}\right)^{\frac{1}{2}}\frac{V^{\frac{3}{2}}}{x^2}$$

where $V$ is the potential at distance $x$ from the cathode.

Hence find an expression for the velocity, $v$, at $x$ and so show that the transit time $\tau$ from cathode to anode is given by

$$\tau = \frac{3d}{(2e/m)^{\frac{1}{2}}V_a^{\frac{1}{2}}}$$

where $d$ is the electrode spacing and $V_a$ is the anode voltage.

Discuss the significance of the result. (University of Sheffield, 1967.)

*Solution.* The first part of the question is covered in §6.4.2. Put

$$\frac{4\epsilon_0}{9}\left(\frac{2e}{m}\right)^{\frac{1}{2}} = k$$

then
$$V = (J/k)^{\frac{2}{3}} x^{\frac{4}{3}} \tag{10.47}$$

Also
$$Ve = \tfrac{1}{2}mv^2$$

so
$$v = (Ve/m)^{\frac{1}{2}} = (2e/m)^{\frac{1}{2}} (J/k)^{\frac{1}{3}} x^{\frac{2}{3}}$$

The time taken for an electron to travel a distance $dx$ is $dx/v$, so

$$\tau = (2e/m)^{\frac{1}{2}} (J/k)^{-\frac{1}{3}} \int_0^d \frac{dx}{x^{\frac{2}{3}}} = (2e/m)^{\frac{1}{2}} (J/k)^{-\frac{1}{3}}\, 3d^{\frac{1}{3}}$$

Substituting for $J/k$ from (10.47), we find

$$\tau = 3d\,(2e/m)^{-\frac{1}{2}}\, V_a^{-\frac{1}{2}}$$

This result is of importance in the design of thermionic valves to operate at very high frequencies. For efficient control of the electron stream, the transit time between cathode and grid must be small compared with a period of the oscillation.

2. In the simple mass spectrometer represented in fig. 10.16, a mixture of singly charged ions, which have been accelerated from rest through a potential difference, $V$, is injected as a parallel beam through the slit $S_1$ into a uniform magnetic field of density $B$. The beam emerging through the slit $S_2$ is collected.

How would you determine which ions are present in the injected beam?

Obtain an expression for the radius $r$ of an ion trajectory, and hence propose suitable values of $B$, $V$ and the widths of slits $S_1$ and $S_2$ for a small spectrometer ($r = 1$ cm) in which ions of mass 4 to 44 are to be distinguished unambiguously from one another.

What would be the effect if (*a*) the injected beam diverged from the slit $S_1$ with an angular spread of $2\alpha$ instead of being strictly parallel; (*b*) doubly-charged ions were present?

(For protons $e/m = 0.957 \times 10^8$ C kg$^{-1}$.) (University of London, 1969.)

*Solution.* If $v$ is the velocity of an ion passing through $S_1$

$$Ve = \tfrac{1}{2}mv^2 \quad \text{and} \quad mv^2/r = Bev$$

where
$$r = (2mVe)^{\frac{1}{2}}/B$$

$B$ should be of such a value that the field can be provided by a permanent magnet and, as a first trial, we put $B = 0.5$ T. We then find, from the above equation

$V = 1196$ for a proton

$V = 299$ for an ion of mass number 4

$V = 27.19$ for an ion of mass number 44

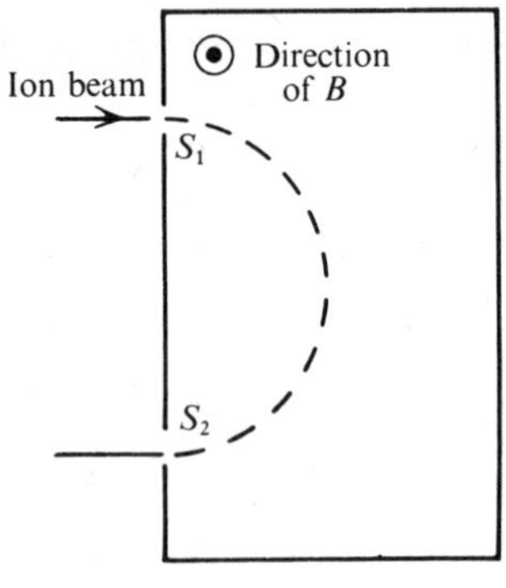


Fig. 10.16

These are reasonable values and the presence of any ion will be indicated by the value of $V$ needed to ensure collection. We therefore take $B = 0.5$ T.

To distinguish between ions of adjacent mass number will be most difficult at the top end of the mass scale, where the percentage change for one mass unit is least. The width of slit must therefore be chosen to distinguish mass 43 from mass 44. $S_1$ determines the width of the ion beam and discrimination will be greatest when $S_2$ is just wide enough to accept the whole beam. We therefore take the widths of $S_1$ and $S_2$ to be equal.

With $V = 27.19$, the diameter of the trajectory is 2 cm for mass number 44 and 1.9772 cm for mass number 43. Thus a slit width of 0.02 cm should ensure complete discrimination between these ions.

If the ions enter the magnetic field with an angular spread of $2\alpha$, they will still have circular trajectories with radius $r$ but, from simple geometry, the slit separation for best collection of the extreme rays should be $2r \cos \alpha$ instead of $2r$. With a slit width of 0.02 cm for $S_2$, $2r \cos \alpha$ should not differ from $2r$ by more than, say, 0.002 cm. Thus $\alpha$ should not exceed about 2.5°. To offset the spread of the beam, some reduction in the width of $S_1$ might be worth while.

A doubly charged ion of mass $m$ will be collected with the same value of $V$ as a singly charged ion of mass $m/2$.

## 10.7 Problems

1. A charged particle, which has been accelerated from rest by passage through a potential difference $V_0$, enters the field between two large plane horizontal conductors through a small hole in the lower conductor. The distance between the conductors is $d$ and the potential difference between them is $V$. The sign of $V$ is such that the force on the particle is directed towards the lower plate. At the point of entry, the particle is moving in a direction making an angle of 45° with the normal to the plates.

Show that, if the particle strikes the upper plate, its horizontal distance $x_1$ from the point of entry will be given by

$$x_1 = V_0 d\{1 - \sqrt{[1-(2V/V_0)]}\}/V$$

If, on the other hand, it strikes the lower plate, the distance $x_2$ will be

$$x_2 = 2V_0 d/V$$

N.B. These results are independent of the charge and mass of the particle.

2. Conditions are the same as in the previous example except that, at the point of entry, the particle is moving in a direction making angle $\theta$ with the normal to the plates. $V$ is large enough to ensure that the particle strikes the lower plate at distance $x_2$ from the point of entry.

It is desired that particles moving in the same plane at the point of entry, but having slightly different values of $\theta$, should have nearly the same value of $x_2$. What would be the best mean value $\theta_0$ of $\theta$ to achieve this result?

Using this value $\theta_0$, by how much could $\theta$ differ from $\theta_0$ if the variation of $x_2$ from its mean value were limited to 2 per cent?

3. An electron moves with velocity $v$ at right angles to a uniform magnetic field of flux density $B$, and so describes a circular orbit. Show that the frequency $f$ with which it completes an orbit is independent of $v$, so long as relativity corrections are negligible, and find the value of $f$.

$f$ is known as the cyclotron frequency.

4. An electron starts from rest in crossed uniform electrostatic ($x$-axis) and magnetic ($y$-axis) fields. Show that, in the subsequent motion of the electron, its maximum displacement in the $x$-direction is $2Em/B^2e$.

# Answers to problems

## Chapter 2

1. No current can cross an edge, which must therefore be a flow line.
2. $\boldsymbol{A} = 4\boldsymbol{i}+1\boldsymbol{j}-2\boldsymbol{k}$, $\boldsymbol{B} = -1\boldsymbol{i}+3\boldsymbol{j}-5\boldsymbol{k}$.
$A = 4.58$, $B = 5.92$, $\boldsymbol{A}\cdot\boldsymbol{B} = 9$, $\theta = 70.6°$.
3. 27.3 $\mu$A.
4. 63.7 $\Omega$. (*a*) 79.5 V, (*b*) 7.95 V.

## Chapter 3

3. $11Q/500\pi\epsilon_0$.
4. Potential with reference to a point at infinity is infinite. Thus the reference must be to some point at an arbitrary distance from the axis.
5. Let $E_1$ be the field due to the conductor whether just outside or just inside the hole. Now suppose the hole filled by a metal disc with the same charge density $\sigma$. This will not alter $E_1$, but will produce an additional field $E_2$ outside and $-E_2$ inside. We now have a closed conductor, so

$$E_1+E_2 = \sigma/\epsilon_0, \quad E_1-E_2 = 0.$$

6. (*a*) $3Q(y-x)/4\pi\epsilon_0 xy$, (*b*) $Q/4\pi\epsilon_0 z^2$.
7. The field will be concentrated near the sharply curved inner conductor and will not be greatly affected by the shape of the outer conductor. It is therefore reasonable to assume that the capacitance per unit length will lie between two values for coaxial cylinders, (*a*) with an external radius of 5 cm, (*b*) with an external radius of $5\sqrt{2}$ cm. These values are (*a*) 14.22 pF, (*b*) 13.06 pF, with a mean of 13.64 pF.

## Chapter 4

1. $10\boldsymbol{i}+3\boldsymbol{j}+11\boldsymbol{k}$.
6. $2\times10^4 \ln 2$ Wb.
9. 52.5 m s$^{-1}$; 69.4 m s$^{-1}$.

## Chapter 5

1. 0.545 $\mu$C m$^{-2}$; $\theta = 66.59°$.
2. Initially the field strength in air will be increased, so breakdown will occur. This will probably cause build-up of charge on the glass, which will reduce the field in air until it falls below the breakdown value.

3. We have two cylindrical capacitors in series. The intermediate conductor is insulated so the charge on it must be zero. Increasing $L$ reduces fields in the inner capacitor and increases those in the outer, so $L$ should be chosen to equalize the maximum fields in the two capacitors. These two conditions determine the ratio of $L$ to $D$. The absolute value of $D$ can then be chosen to make the fields a minimum.

## Chapter 6

1. $E = 3V_0/8R$.
4. The two image charges are of equal magnitude and opposite sign, but the one producing an attractive force is nearer to $q$ than the other. Thus, their combined result is an attractive force proportional to $q^2$. If $q$ is large enough, this can always be made larger than the repulsive force due to $Q$, which is proportional to $Qq$.
5. Since the plates are of infinite extent, their capacitance to earth is also infinite. Thus the finite image charges will cause no change in the potential of the plates relative to earth and the field will be the same as if the plates were earthed.
Another way of putting this is to say that a charge $+Q$, to compensate the images, would be spread uniformly over an infinite area and would thus produce no appreciable field.
7. $\sigma = 16Q/75\pi a$.
8. The image charges will be $-aq/b$ at a distance $a^2/b$ from the centre $O$ and a compensating charge $+aq/b$ at $O$. In order that these shall make the sphere an equipotential, we have the relation $aq/b \cdot BP = q/AP$, which allows us to express $BP$ in terms of $AP$. The components of field at $P$ arising from each of the charges are readily obtained. They can be further resolved into components along $OA$ and components along $OP$, using $OPB$ and $OPA$ as triangles of forces. Thus the total radial field at $P$ can be found and hence the surface density of charge.
10. Resistance $= 11t/8$ approximately.

## Chapter 7

1. Flux densities (*a*) 1.33 T, (*b*) 1.66 T.
Number of turns (*a*) 1200, (*b*) 1400.
2. Reluctances (*a*) $5.03 \times 10^5$, (*b*) $1.52 \times 10^5$.
Flux densities (*a*) 0.5 T, (*b*) 0.39 T.
3. 1.37 T.
4. 17.2 cm; 1.86 T.

## Chapter 8

1. The battery is better by a factor of about 1000.
2. $\epsilon_0(\epsilon_r - 1)V^2/2d^2$.
3. Force (attractive) $= 3\pi\mu_0 R_1^2 R_2^2 N_1 N_2 I_1 I_2 x/2(R_1^2 + x^2)^{\frac{5}{2}}$.

4. $IAB \sin \theta$.

6. $f(\theta) \propto \sqrt{\theta}$; yes, for any given value of $\theta$, the torque is still proportional to $I^2$.

## Chapter 10

2. $\theta_0 = 45°$; $\delta\theta = 5.7°$.

3. $Be/2\pi m$.

# Index